The Primer Fields

by Rolf A. F. Witzsche

Contents

About the Illustrated Science series
*On the Ice Age and Climate Change
and the book*

The Primer Fields

Book 1 of the series: Ice Age of the Dimmer Sun in 30 Years

The term, 'Primer Fields' refers to electromagnetic structures in space that focus interstellar plasma in concentrated form onto our Sun. Without the Primer Fields the Earth would be in an Ice Age environment. We, ourselves, would likely not exist then. The fields are created by magnetic effects of flowing plasma acting on itself. A threshold must be exceeded for the fields to form. Below it, the fields cannot form. The Sun is powered at a lower level when the Primer Fields are not active. At the inactive state the solar activity is reduced to a type of cosmic default level with 70% less radiated energy. The phase shift starts an Ice Age. At the present rate of diminishment, the solar activity threshold may be reached in 30 years, or in the 2050s, most likely. That's when the interglacial period ends and the greatest Climate Change in recorded history, happens.

With the primer system gone inactive, the climate on Earth will get 40 times colder than the Little Ice Age in the 1600s had been. Ice core evidence promise that. Without the needed preparations for human living in such an environment, 99% of humanity would die of starvation, both by the cold and by CO_2 depletion as more of it becomes dissolved into the sea, that severely affects agriculture.

With the Primer Fields being critical for our very existence, they have been explored in laboratory experiments. Only their absence has not been explored, which needs to be explored before it happens, while a preparatory response is still possible.

Numerous fields of evidence tell us that the next Ice Age is near. That's where the truth begins. Most of the evidence was discovered in the 1990s and thereafter. Some evidence is measured in ice cores; some is measured in space, by satellites. Some measurements are also made on the ground in terms of measurements of the Earth's magnetic-pole drift observed in northern Canada. All of this is seen combined with high-energy physics experiments at a leading national laboratory, and is also explored in the small in static experiments.

 Against the background of these widely diverse types of evidence that have been recently discovered, the historic Little Ice Age in the 1600s, takes on a new dimension as a yardstick for measuring the future that by this evidence promises to be up to 40-times colder than the Little Ice Age had been. It qualifies for the term, Absolute! The evidence poses a great challenge ahead. Are we ready to respond? The Ice Age phase shift in climate is a stark in differences as night and day, and similarly fast.

In the Little Ice Age between 10% and up to 30% of the populations in Europe had perished by starvation. The last Big Ice Age was evidently vastly harsher. Only 1-10 million people emerged from it alive. That's all we had after 2 million years of development. We want to do far better this time around; and we can, with large-scale technological infrastructures for our food supply. But will we create them? Will we get the job done in the 30 years that we still have left before the Ice Age starts anew? Will we even consider it? And how certain are we that the phase shift to the next glaciation period will begin, as the evidence suggests, in the 2050s? We have no slack on this front. We have no slack on this front. Should we fail us on this absolute front, we would be committing suicide.

So, what will the answer be? Will we move with the evidence? Or will we lay ourselves down to die by default?

It takes an independent researcher to brake the taboos that have kept mainstream cosmology imprisoned, increasingly, during the past century, even while what is regarded as taboo is known to be wrong.

The Illustrated Science series is intended to open the scene beyond the threshold of accepted taboos, to where the actual physical evidence speaks for itself.

The scope of the existential challenge that the Ice Age brings with it, takes astrophysics out of the academic domain and places it into the foreground as one of the most-critical issues of our time. The big Climate Change events that have already worldwide effects are mere fringe effects in the flow of the ever-changing cosmic dynamics. The big effect, when the Ice Age begins anew, promises to be caused by a dimmer and colder Sun with 70% less radiated energy. This defines our climate future.

Sure, we can live with all that by creating new platforms for agriculture that are able to operate under Ice Age conditions. But will we do it? The task is enormous. Or will we fail ourselves on this front? We have no reason to allow us to fail. We have the materials and energy resources on hand to accomplish everything that is required for us to continue to live in an Ice Age World. But will we do it? The big question that never goes away, therefore, is; will we develop our inner resources as human beings sufficiently to get the job done, and to get it done in time? Or will we do nothing, ignore the challenge, and condemn our children and one-another to an agonizing death by starvation? That's the choice.

Towards meeting the inner challenge, I have created the epic series of novels, The Lodging for the Rose. And further, towards meeting the science challenge, I have produced numerous research books and several dozen exploration videos that the Illustrated Science series is modeled after. The work is the result of a quarter century of research, for which numerous elements of evidence in related fields came to light during the timeframe of my research.

It is my hope that the work that went into all of these projects will help in some degree - for humanity that we are all a part of - to write itself a ticket to have a future.

High-resolution color images, of the images in this book, can be obtained at www.iceagetheatre.ca

Part 1
Introduction to the dimming Sun

Transcripts at: www.ice-age-ahead-iaa.ca

Part 1
The Primer Fields

Introduction to the dimmer Sun
By popular request, I have updated and reproduced my video, 'Ice Age of the Dimmer Sun in 30 Years,' with a new voice, a new form, with leading-edge discoveries in science added, and with a new musical theme. And to make the content more accessible, I have divided the video's ten parts into a series of ten individual videos.

The science-focus of the series

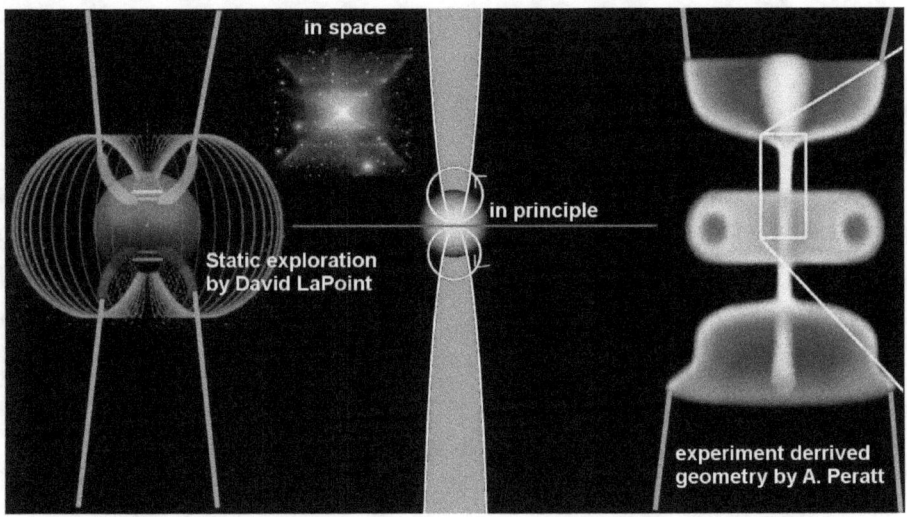

The science-focus of the series, of course remains the same as in the original video, being centered on the Primer Fields that are understood as one of the most fundamental aspects of the science of the universe, the galaxies, the solar system, the Sun, and the effects of the Sun on the Earth, ranging from the color of the sunlight to solar cosmic-ray flux affecting our climate that we enjoy and also fear.

So what are these all-pervading Primer Fields then, that seem to be affecting everything?

The Primer Fields

David LaPoint - The Primer Fields

In December 2012 the plasma physicist David LaPoint published a series of videos about a phenomenon that he called: 'The Primer Fields.' With his discovery he takes aim at long-held mistaken theories about the physical universe, including black holes, dark matter, and dark energy, which are not required to interpret the observed evidence of the universe.

The most basic electromagnetic fields

The Sun at
a plasma node

David LaPoint - The Primer Fields

In his videos David LaPoint illustrates the existence, and the operating principles, of the most basic electromagnetic fields that shape the solar system, the galactic system, and ultimately the cosmic environment. He suggests that these basic primal fields, which he discusses, are fundamental to the order and operation of the universe, and are even reflected in the transmission of light and the structure of matter itself, all the way down to its smallest subatomic forms.

The Primer Fields and their effects on our Sun

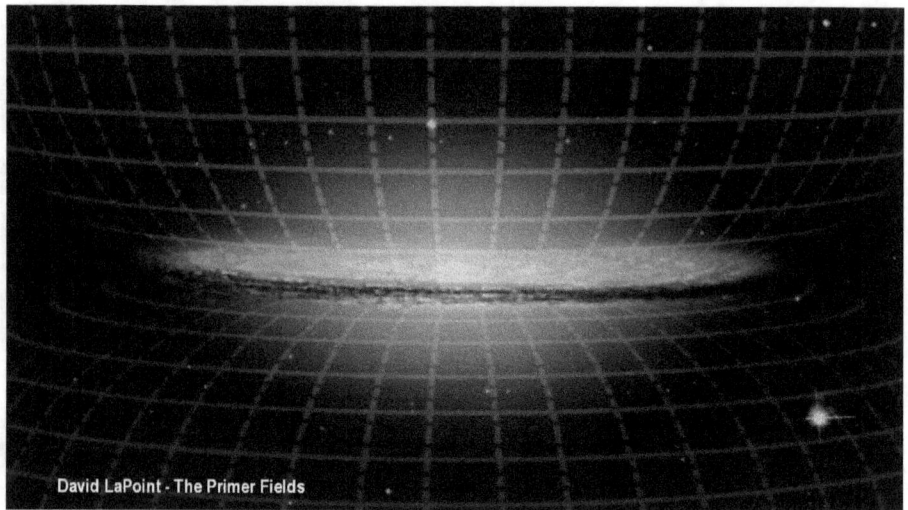

David LaPoint - The Primer Fields

On the gigantic scale the electromagnetic fields that he recognized to exist, and has explored in the laboratory, shape entire galaxies. They are so fundamental to almost everything, that he calls them "The Primer Fields." In fact, without the Primer Fields and their effects on our Sun and planets, even the galaxy itself, might not have been created, so that nothing would actually exist without them.

Intensely focused and compressed plasma

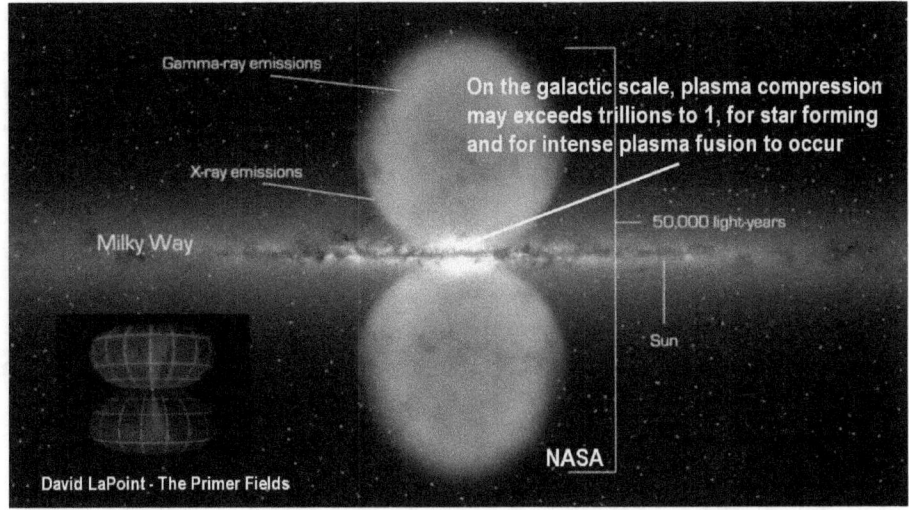

Gamma-ray emissions

X-ray emissions

Milky Way

On the galactic scale, plasma compression
may exceeds trillions to 1, for star forming
and for intense plasma fusion to occur

50,000 light-years

Sun

NASA

David LaPoint - The Primer Fields

Star formation in a galaxy typically occurs in intensely focused and
compressed plasma that the Primer Fields facilitate, to the point
that nuclear fusion occurs, by which atoms, stars, planets, and
entire galaxies are formed.

Our solar system as a single functional unit

David LaPoint - The Primer Fields

The term, Primer Fields, is also justified, because on the ' smaller' stage of our solar system, the operation of the Primer Fields, though they are correspondingly smaller in size than on the galactic scale seen here, is critical for the dynamic operation of our solar system as a single functional unit.

The dynamics of the ice ages

A plasma sun
born in the
laboratory

David LaPoint - The Primer Fields

David LaPoint also discovered that the principles of the fields, that he calls the Primer Fields, is such that the galactic plasma streams that flow through the solar system become concentrated to a high degree of density around the Sun, which he replicated in principle in laboratory experiments. The result of his work brings a radically new dimension to the perception of the Sun and also of the dynamics of the ice ages.

*The Sun as an electrically powered star

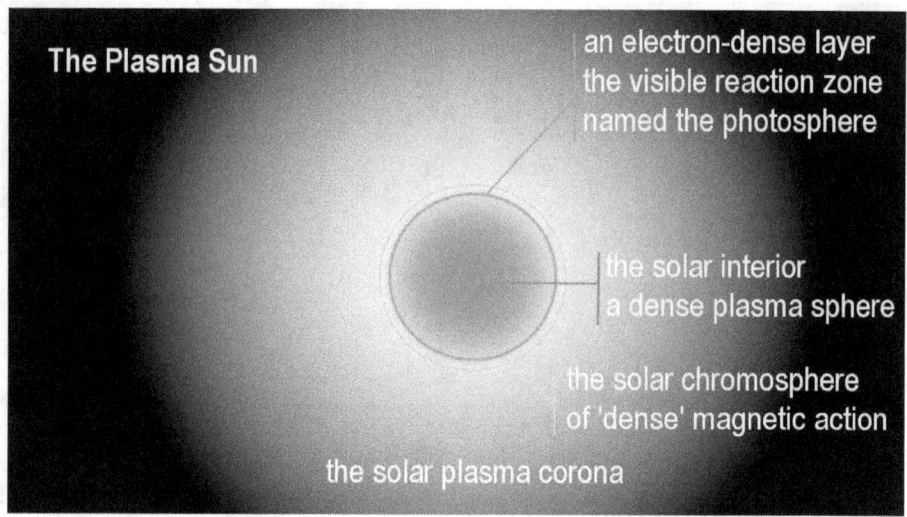

The Plasma Sun

an electron-dense layer
the visible reaction zone
named the photosphere

the solar interior
a dense plasma sphere

the solar chromosphere
of 'dense' magnetic action

the solar plasma corona

The perception of the Sun as an electrically powered star has existed for a long time already. A large body of evidence exists to support the electric Sun theory. But it had a weak point. For the theory to work, a high concentration of electric plasma is required to exist around the Sun, which put the theory in doubt. As an alternative, the theory of the nuclear fusion-powered Sun was created, except the observed evidence doesn't support this theory either.

Sunspots on the surface of the Sun

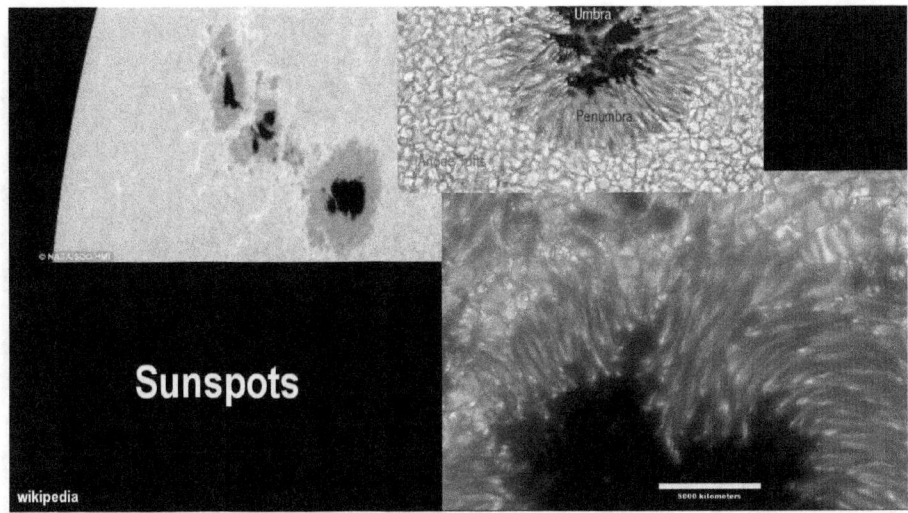

The sunspots on the surface of the Sun show the Sun to be darker below the surface. We should see the opposite if the Sun was internally powered, instead of it being electrically powered at the surface.

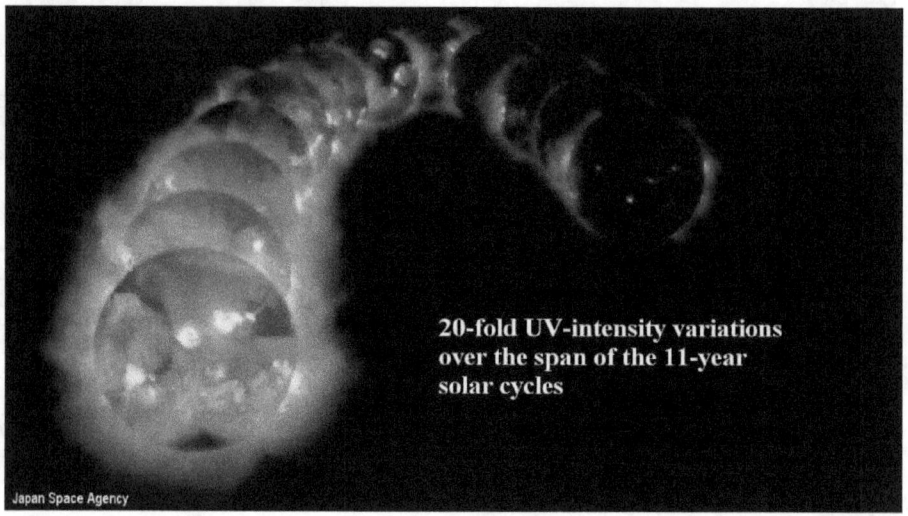

20-fold UV-intensity variations over the span of the 11-year solar cycles

Japan Space Agency

Also the Internal-Nuclear-Fusion-Sun theory cannot readily explain the 11-year solar activity cycles in which the brightness of the Sun varies by a factor of 20 when observed in the high ultraviolet band. The fusion Sun theory has it that the photon travel time from the Sun's deemed fusion center to the surface, is deemed to be in the range between 10,000 and 170,000 years, while the assumed energy-transfer time itself, from the center of the sun to the surface, is deemed to be on the order of 30 million years. This hardly supports the 11-year solar cycles, in which the Sun's magnetic field changes direction at every cycle.

A minimal threshold of conditions

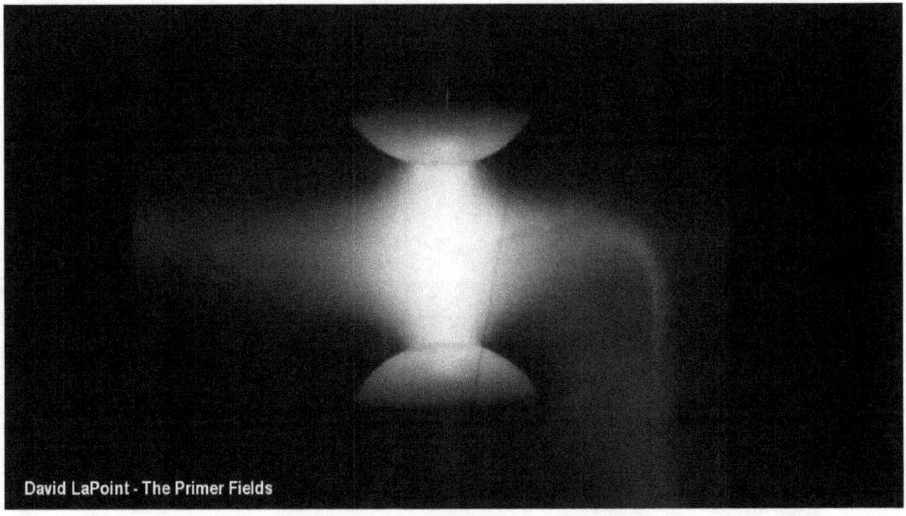

David LaPoint - The Primer Fields

The contribution that David LaPoint brings to the table in this debate weighs enormously in support of the Electric-Sun theory by illustrating how the high density plasma field is created around the Sun that the theory depends on. The Primer Fields theory takes away the barrier against the Electric-Sun theory, which thereby becomes totally plausible, together with a number of related theories that thereby likewise become plausible. However, the Primer Fields theory takes us one critical step beyond merely supporting the Electric-Sun theory. It renders the functioning of the electric Sun subject to a minimal threshold of conditions that must be met for the Sun to be powered, which, when it is not met, causes the Sun to become inactive, dim, and 'cold.'

If the threshold is not reached

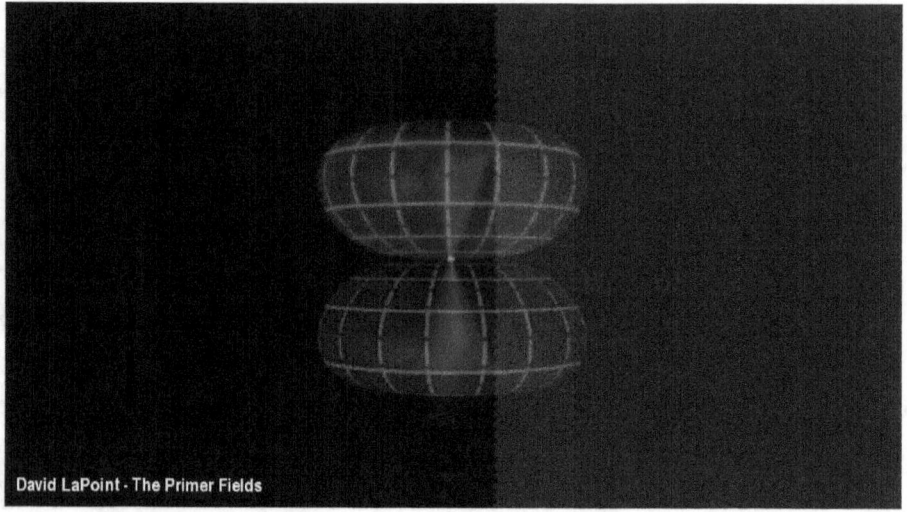

David LaPoint - The Primer Fields

While David LaPoint may not have intended to support the electric-sun theory, his work illustrates that the process that creates a dense plasma sphere around the Sun happens, because the Primer Fields have that effect. It is known in plasma physics that these fields themselves depend for their existence on a minimal density in the plasma that flows through the system. If the threshold is not reached, the fields do not form. If the fields do not form, or cannot be maintained, the Sun does not have the condition established for it to be powered.

The Sun simply turns off

The Sun simply turns off to an inactive state in which it glows dimly by its stored up energy and some nuclear decay processes within it.

When our brilliant sun is turned off

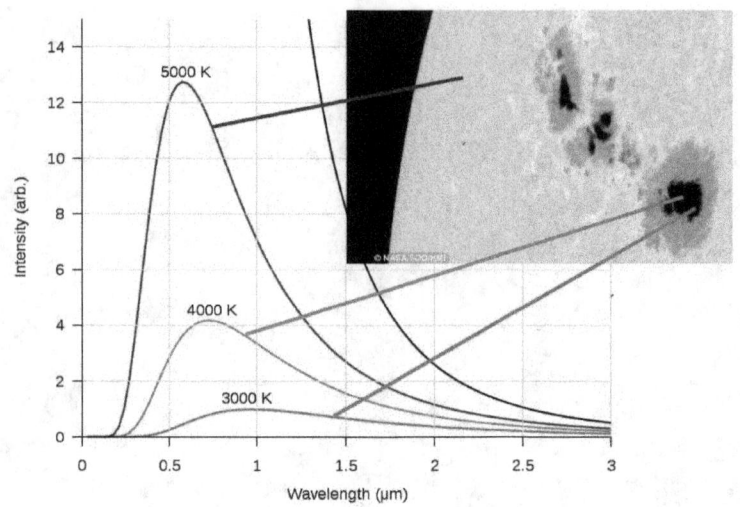

The required threshold, therefore, becomes an important factor in considering the dynamics of the ice ages. Just imagine the consequences on the Earth when our brilliant sun is suddenly turned off as the Primer Fields collapse, and looses more than two-thirds of its energy output in possibly a single day.

The Milankovitch theory

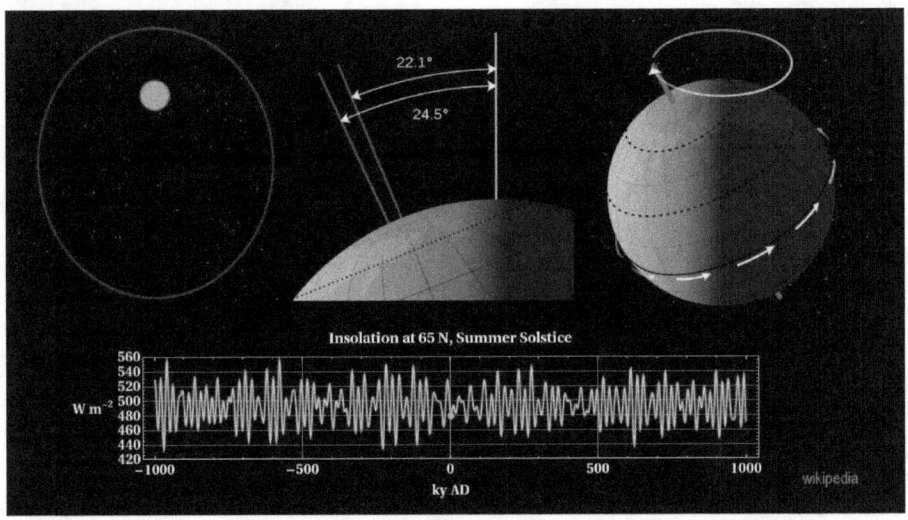

The conventional theories of the ice age dynamics all regard the ice age glaciation as the result of a gradual cooling of the Earth. One theory sees the cooling being caused by cyclical variations of the orbit of the Earth around the Sun, with the Sun remaining an invariable constant by this theory. In the Fusion-Sun theory, the Sun is deemed to be an invariable constant, so that the ice ages are caused by anything else except the Sun.

The resulting orbital-variation theory, known as the Milankovitch theory, however has so many holes in its fabric that it is generally no longer seriously considered as a cause for the ice ages.

The large temperature fluctuations

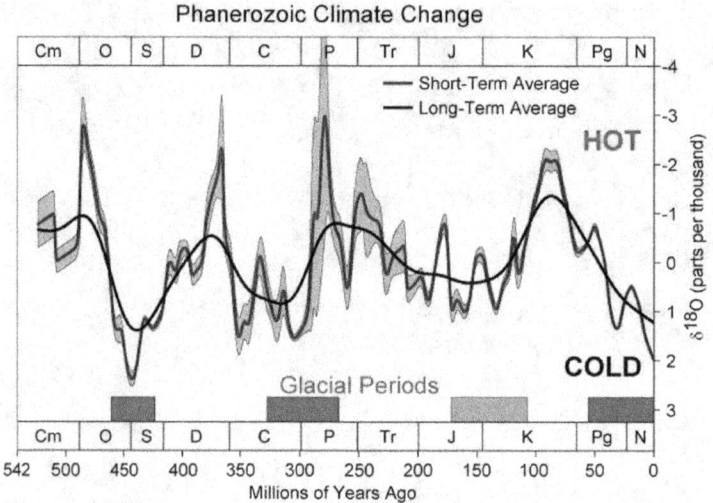

Phanerozoic Climate Change

The standard electric-sun theory doesn't have this problem. Plenty of evidence exists that the power output of the Sun does vary with the density of the electric input. The large temperature fluctuations that are known to have occurred over the last half billion years are evidently the result of cyclical electric input fluctuations. The Primer Fields theory does not change that, but it does open the gate to the recognition that the entire process of the powered Sun can collapse when a minimal power-input threshold is not met, as we have seen happening during the ice ages.

Primer Field theory brings a huge difference

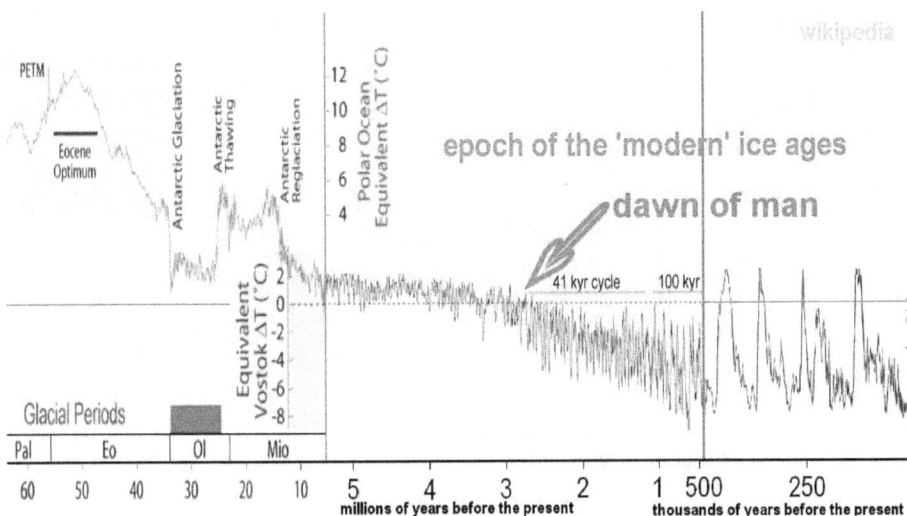

The Primer Field theory brings a huge difference to the table. In the past it was believed that ice age conditions develop slowly and gradually over extended periods, so that humanity can adjust itself to the changing conditions, even in cases when the Earth becomes five times colder than the Little Ice Age had been, over the span of 50 years. The Primer Fields theory jolts us out of this dream, or at least it should.

The Primer Fields theory presents us a great blessing

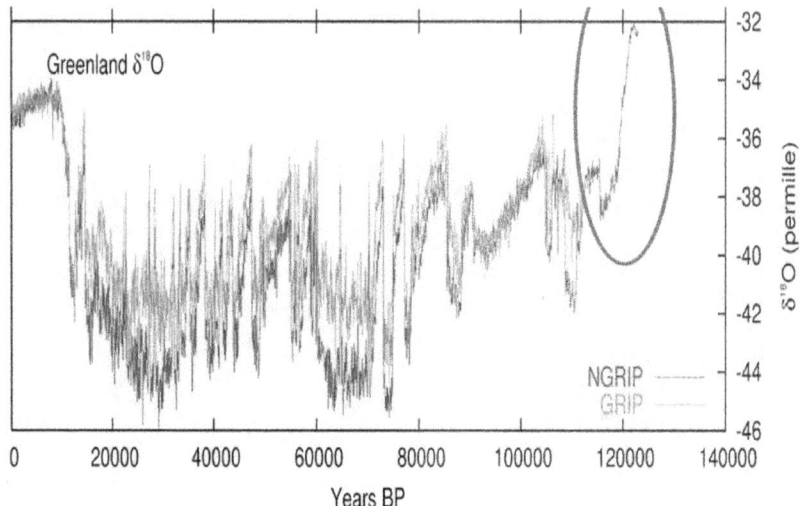

If the fields are not maintained sufficiently that they collapse and vanish, extremely radical changes occur in as short a time as possibly a single day. The Primer Fields theory takes away the general notion that we have a long time to prepare for the coming Ice Age and don't need to respond yet. It tells us that the opposite is true; that we need to get prepared as fast as possible, which almost no one is yet willing to even consider. In this context the Primer Fields theory presents us a great blessing in disguise.

To avoid the catastrophe of nuclear war

USAF B-2 Spirit
Stealth bomber

20 in service
9,700 Km range without refuelling
Bomb capacity app. 38,000 pounds
Price tag: $2,000,000,000

wikipedia

It forces us, if we are willing to respond, to re-develop our
humanity as fast as possible, which we urgently need in the present
to avoid the catastrophe of nuclear war that no one will likely
survive, for which the preparations are evermore intensely pursued
in numerous ways.

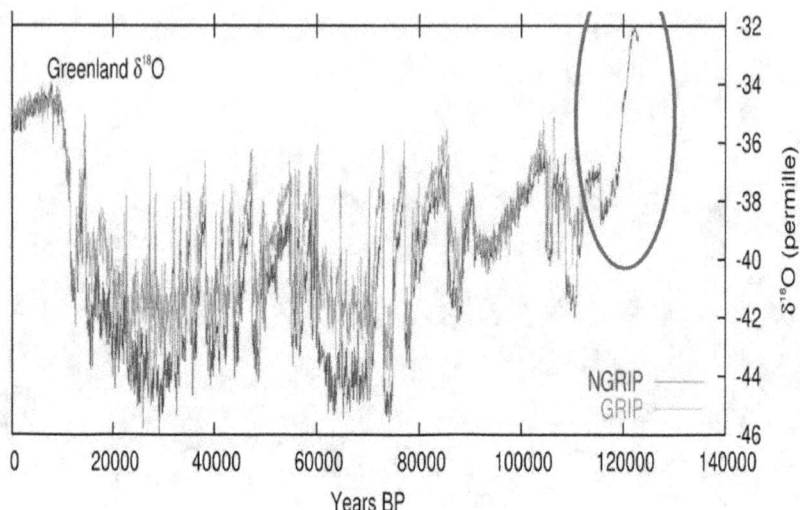

In this regard the potentially near, rapid ice age transition that the Primer Fields theory brings into view, offers us an incentive to get our act together, which would be the greatest salvation that we could experience on every front where civilization is presently fast collapsing.

A single basic principle

The Primer Fields theory illustrates a single basic principle that applies to all complex systems, including civilization. In economics, when the creative and productive power of a nation diminishes below a minimal threshold, the entire economic system disintegrates. It becomes inactive. The nations collapse. The world becomes cold. The people die.

Close to the minimal threshold

Annihilation is assured

500,000 times
Hiroshima
in one hour

Castle Bravo - the first U.S. test of a dry fuel thermonuclear hydrogen bomb - March 1, 1954 at Bikini Atoll, Marshall Islands

This happens to cultures too when the recognized value of our humanity drops below a minimal threshold. Nuclear war happens then, when the humanist culture disintegrates.
We are close to the minimal threshold on a number of these vital fronts, especially in economics.

When the assumed value of money becomes uncertain

The High Five in Dallas, Texas, USA - Wikipedia

For example, when the assumed value of money drops below the point where the value of money becomes uncertain in the markets, it becomes meaningless at this point. At this point the financial markets collapse, banks close, stores close, gas stations close, transportation collapses, the food supply system stops, the entire platform of physical infrastructures becomes inactive.

The Glass Steagall law

1933 - the Glass-Steagall law to protect that value of the U.S. currency - now repealed

The world's money-value system can unravel in a single day. No one knows how close we are to this day. The Glass Steagall law that once protected the value of money in the USA, has been repealed in Congress in 1999 by a bunch of traitors who were bribed to do so with a $350 million slush fund. To date all efforts have failed to restore even this minimal protection of the value of currency that the repealed law had afforded for the 66 years in which it was active. Against the background of the now fast collapsing value of money the entire economic system can collapse in a single day, just as the presently powered Sun can go inactive in a day when the fields collapse that prime its operational environment.

During the interglacial warm period

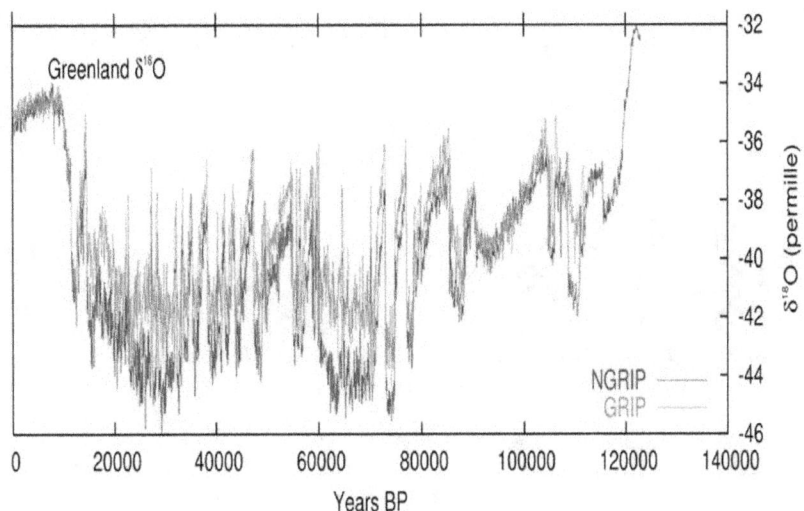

David LaPoint's work illustrates that the electric Sun, which varies only slightly during the interglacial warm period that we are presently enjoying, will with great certainty go completely inactive when the plasma density of the electric streams feeding into the solar system falls below the threshold level where the primer fields collapse.

In astrophysics, that's the point where the ice age transition begins. That's what happened 120,000 years ago.

We will cross the cut-off threshold

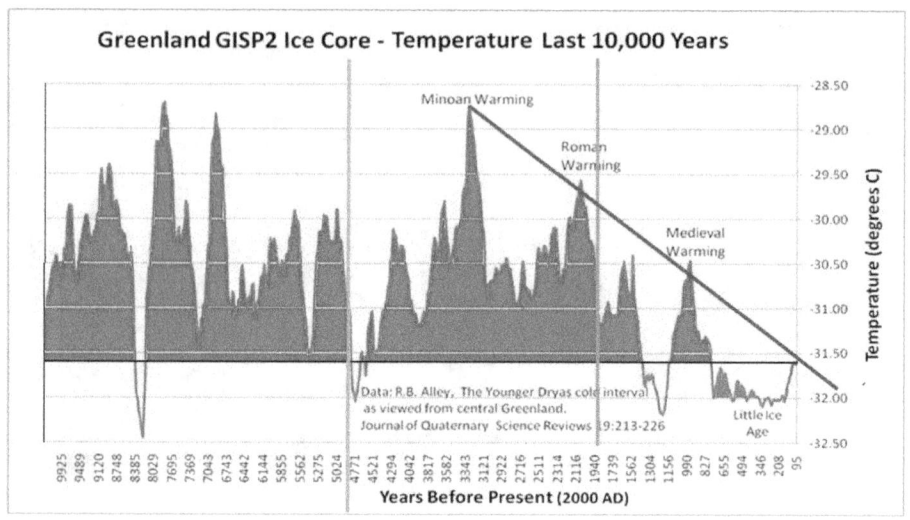

The current trend suggests that we will cross the cut-off threshold in the near future. An ice age begins when the minimal conditions no longer exist for the system to function, that powers the Sun. The Sun reverts to its default state then - its inactive state. It turns dim in a single short step, and the climate on earth that looses two thirds of its energy input, turns cold. At this point rapid deep cooling begins all across the Earth.

Before this point is upon us, all the preparations for the dramatic changes in the Earth's environment will have to be completed. To fail is not an option if we want to continue to live. For this reason, it is not an option either that we raise the value of our humanity far above the present level and protect it on all fronts. And this needs to begin now.

The Sun became inactive 120,000 years ago

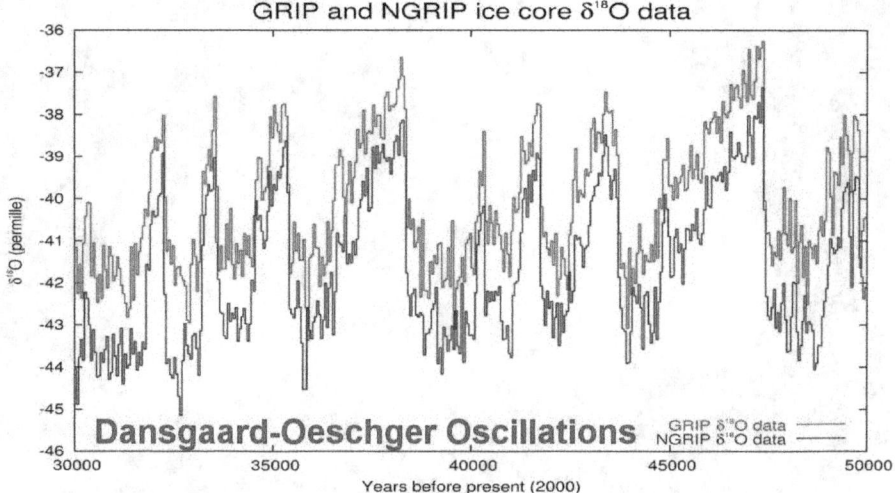

The analysis of ice cores drilled from the ice sheets of Greenland indicates that the Sun became inactive 120,000 years ago. That's when the last Ice Age started. The ice core records also tell us that the Sun became activated for short periods, every 1470 years, throughout the entire glaciation period. The periodic re-warming may have prevented the Earth from freezing up completely.

Ice sheets more than 10,000 feet deep

The inactive periods of the Sun were nevertheless long enough to create the huge cooling on earth that lays up ice sheets across the northern hemisphere more than 10,000 feet deep, which are known to have existed.

The Sun inactive in 30 to 50 years

It is here, that the theoretical becomes important for us all in a critical practical manner. The present trends suggest that the turn-off transition that makes the Sun inactive, might be upon us in 30 to 50 years' time. When this happens the entire world will have to live with a dimmer Sun on a colder earth. Of course this poses some challenges for maintaining our food supply.

The chlorophyll in plants

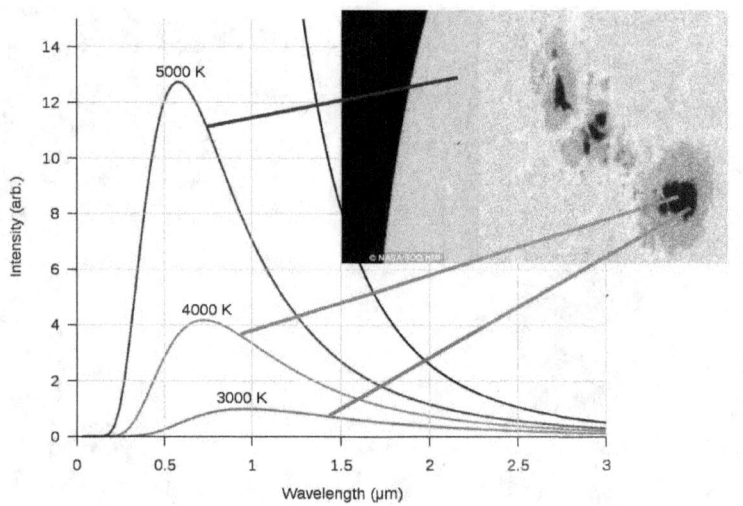

The chlorophyll in plants will have to function at reduced energy levels and with a shifted radiation spectrum.
Here a great challenge arises for humanity as a whole, to work together to create infrastructures that are needed to maintain our food supply in the coming dimmer and colder world.

Yes, we will do this

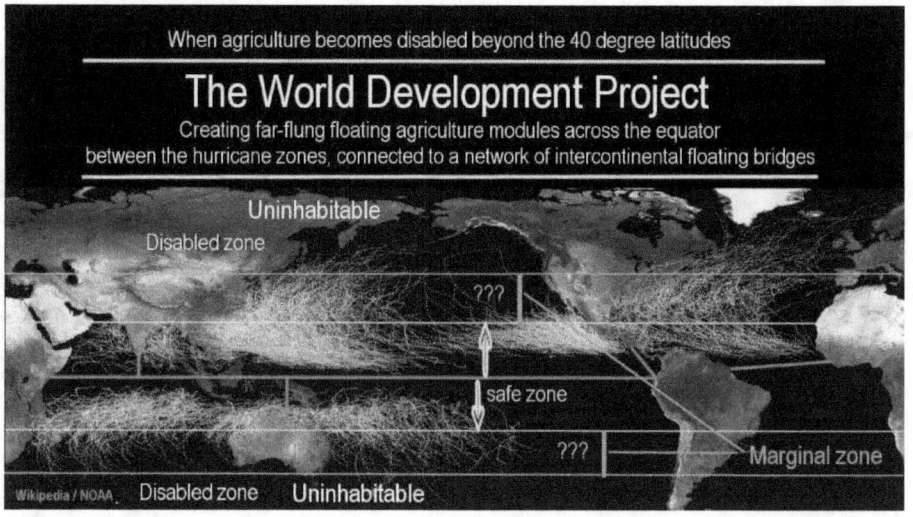

While it is technologically and economically possible to meet the challenge to protect our world, our civilization, and our future, the question remains to be answered whether we will do what is needed to maintain our existence?
The answer must be: "Yes, we will do this. Yes, we will do what is critically necessary for us to survive."

Truthful science supplies the potential

Mass Murder with Biofuels
a YouTube video

El Tres de Mayo, by Francisco de Goya - Wikipedia

We will even do it for far lesser reasons, such as for becoming free of the terrible strangulation of our world that constricted scientific perception has brought upon it. Truthful science supplies the potential to open the door for the liberation of humanity that false science inflicts massively, and globally.

But what is love?

Our physical civilization is largely built on science, first and foremost, just as our social civilization is built on love.
But what is love? We cannot weigh it, measure it, quantify it, but we know that it is so immensely substantial that civilization would collapse without it.
Much of the same can be said of science.

Without science standing at the heart of it

a NASA study on wingtip vortices

s Royce Trent 900 engine for A380 during testing

Without science standing at the heart of it, civilization with all its numerous freedoms would not exist.

False science is destructive

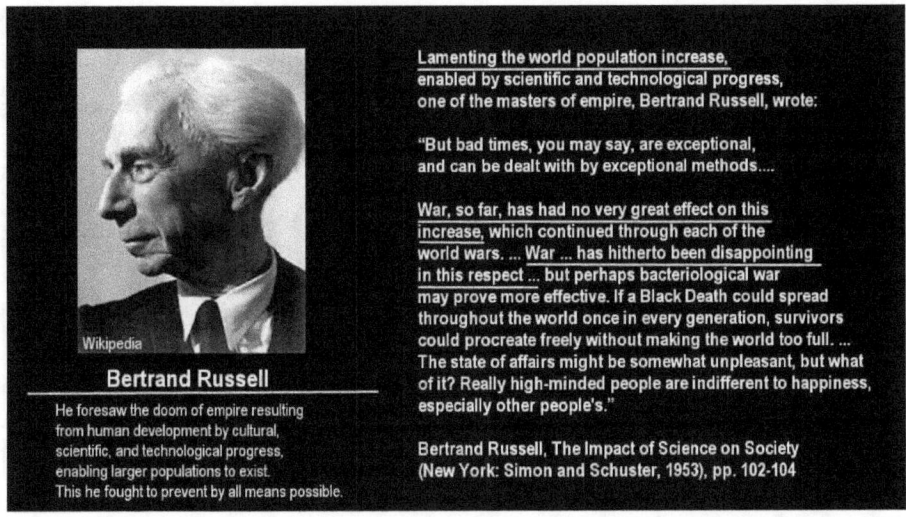

Lamenting the world population increase, enabled by scientific and technological progress, one of the masters of empire, Bertrand Russell, wrote:

"But bad times, you may say, are exceptional, and can be dealt with by exceptional methods....

War, so far, has had no very great effect on this increase, which continued through each of the world wars. ... War ... has hitherto been disappointing in this respect ... but perhaps bacteriological war may prove more effective. If a Black Death could spread throughout the world once in every generation, survivors could procreate freely without making the world too full. ... The state of affairs might be somewhat unpleasant, but what of it? Really high-minded people are indifferent to happiness, especially other people's."

Bertrand Russell, The Impact of Science on Society (New York: Simon and Schuster, 1953), pp. 102-104

Bertrand Russell

Wikipedia

He foresaw the doom of empire resulting from human development by cultural, scientific, and technological progress, enabling larger populations to exist. This he fought to prevent by all means possible.

That false science is destructive is well-recognized by those who would destroy civilization with it. False science has a destructive effect.

Society raising itself out of the trap of false science

Rotor of a steam turbine
by Siemens

With society raising itself out of the trap of false science, a path opens before it to universal liberty, by scientific and technological progress, towards the brightest cultural renaissance of all times. That's not being too optimistic. The potential exists. The principle is applied. It reflects the power of our humanity as human beings.

False science to forcefully strangle civilization

PARIS2015
UN CLIMATE CHANGE CONFERENCE
COP21·CMP11

Poster of the Climate Conference.
Licensed under Fair use via Wikipedia

COP 21: Heads of delegations by GUSTAVO-CAMACHO-GONZALEZ - Licensed under CC BY 2.0 via Commons
by Presidencia de la República Mexicana -delegates

Our present world is full of examples of false science being massively applied to forcefully strangle civilization, and to a large degree voluntarily.
One false science concept is that the amazing recovery of the climate of the Earth from the devastating Little Ice Age in the 1600s, until 1998, was a manmade phenomenon caused by human economy and energy applied for living that must be eliminated. But false science remains false, regardless of how deeply it has penetrated society and controls its responses.

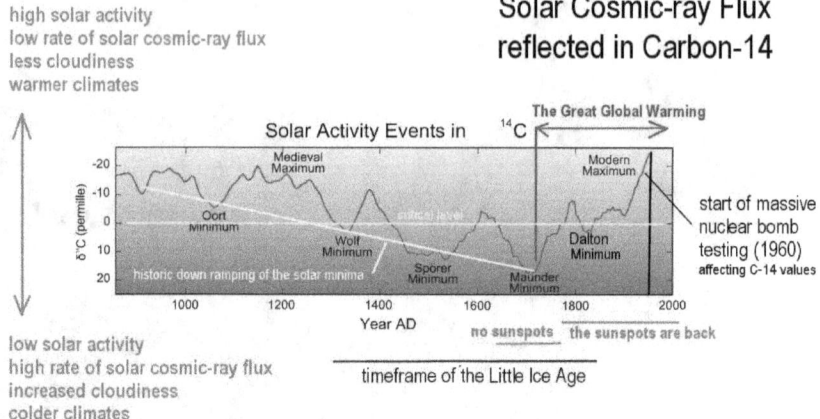

high solar activity
low rate of solar cosmic-ray flux
less cloudiness
warmer climates

Solar Cosmic-ray Flux reflected in Carbon-14

Solar Activity Events in ¹⁴C

The Great Global Warming

Medieval Maximum
Modern Maximum
Oort Minimum
Wolf Minimum
Sporer Minimum
Dalton Minimum
Maunder Minimum

start of massive nuclear bomb testing (1960) affecting C-14 values

historic down ramping of the solar minima

low solar activity
high rate of solar cosmic-ray flux
increased cloudiness
colder climates

no sunspots   the sunspots are back

timeframe of the Little Ice Age

"Carbon14 with activity labels" by Leland McInnes at the English language Wikipedia. Licensed under CC BY-SA 3.0 via Commons

Unrestricted science demonstrates with recognizable certainty that the re-warming of the Earth from its ice house condition in the 1600s was not a human achievement, was not a consequence of mankind's industrial development and its burning of carbon fuels, but was absolutely, and measurably, and scientifically demonstrably, a solar-forced climate effect that has staged the Great Global Warming that enabled modern society and its efficient agriculture to become possible. When honest science displaces constricted science, a great liberation becomes potentially possible.

The solar-forced Great Global Warming

Rate of Change in Icecap Height (cm/year)

Work of the United States Federal Government (2007)

Melting pots on the Greenland Ice Sheets

In the carbon-14 documented example of the solar-forced Great Global Warming, which comes to light as a matter of basic fact in un-constricted science, the liberating effect of truth lifts the burden from humanity of being a climate villain on the Earth that false science has brought like an indictment against humanity, for which the nations are presently induced to stop the burning of carbon fuels and thereby commit economic suicide.

The Greenland Ice Sheet experienced some melting

Yes, the Greenland Ice Sheet experienced some melting of its ice masses that were built up in the deep cold of the Little Ice Age, with the Arctic's sea ice likewise experiencing some thinning out, as it is shown here. However, one doesn't need to have a Master of Science degree to recognize that such effects are in line with the measured increase in solar activity that supplies our climate in the first place.

The Primer Fields stand in the background to the unfolding scientific recognition of the natural dynamics of the forever ongoing climate changes on Earth, slow as the recognition process may be.

Humanity never was a climate factor

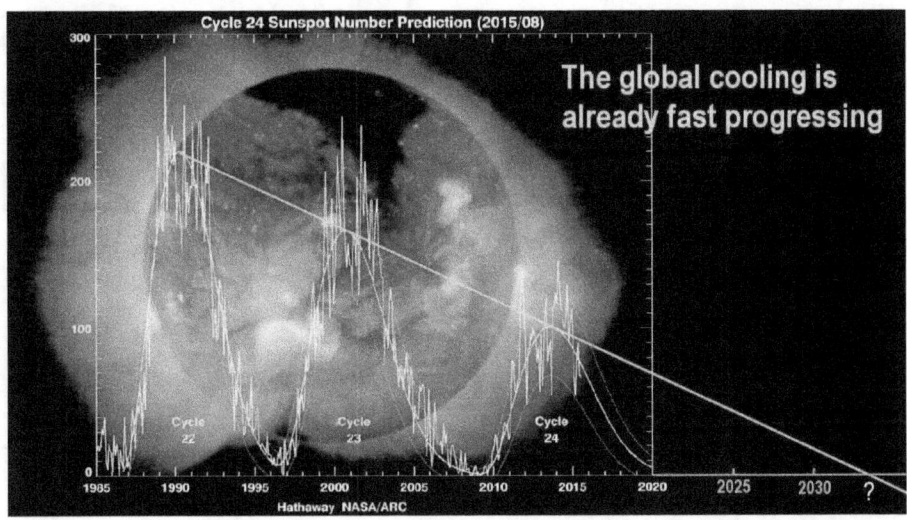

Humanity certainly does not have the capacity, by any means within its grasp, to affect the climate of the Earth. The Earth's climate changes with the effects forced on it by the dynamics of the Sun. Humanity never was a climate factor.

The carbon gases, such as CO_2, that humanity generates by its living, are not a climate factor either - never have been, or ever will be.

The most horrific holocaust

Mass Murder with Biofuels
a YouTube video

El Tres de Mayo, by Francisco de Goya - Wikipedia

With its simple, wide open scientific recognition, honest science presents itself with the potential to liberate humanity from many constricting dogmas, including that for the most horrific holocaust in the history of the world, that is presently ongoing.

In the name of reducing CO_2 emissions by automobiles, vast amounts of agricultural resources are being diverted from the nourishment of people to be burned in the form of biofuels. The burned food resources would nourish upwards to 400 million people, if they were not burned. In a world that has a billion people living in chronic starvation, the food burning unfolds as a holocaust in which 100 million people are forced to die by starvation. The result exceeds the Nazi holocaust 100-fold, and Napoleon's total-war holocaust 1000-fold. Honest science can bring an end to this tragedy that false science has invoked. And here too, the Primer Fields stand in the background to the liberation that simple, truthful, science recognition can inspire.

The electric motivator

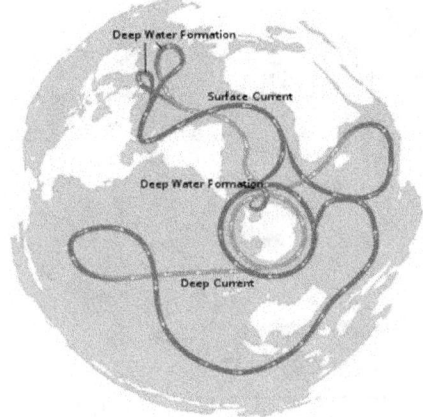

The ocean currents conveyor belt centered on the deep cold waters encircling Antarctica

The Primer Fields stand in this scene as the electric motivator that in part drives the gigantic global seawater recycling system that transports cold, CO_2-rich sea water, from the ice-cold polar oceans that are able to dissolve CO_2 more readily, into the tropical oceans where the cold waters warm up and release the high concentration of CO_2 that rides along in the recycling streams.

The cold deep currents flow slowly

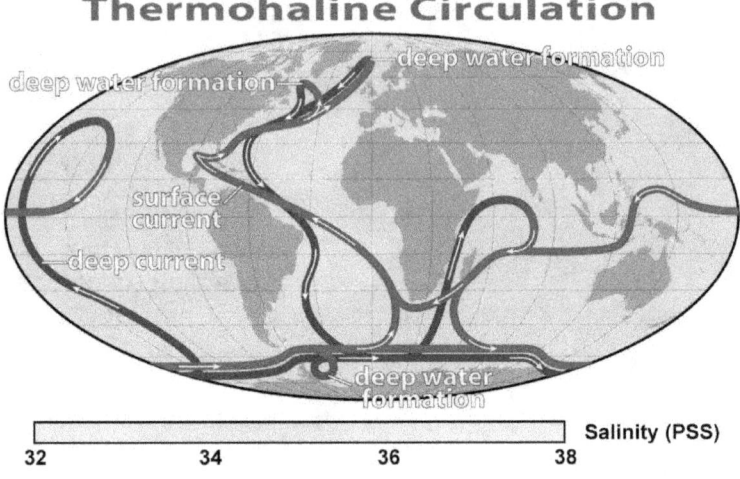

The cold deep currents flow slowly. They flow so slowly that the dissolved CO2 from Antarctica will remain in transit for roughly 350 years until it becomes released back into the air in the tropical oceans near Africa. For the waters from the Arctic, the recycle time likely exceeds a thousand years.

CO2 is coming back to us

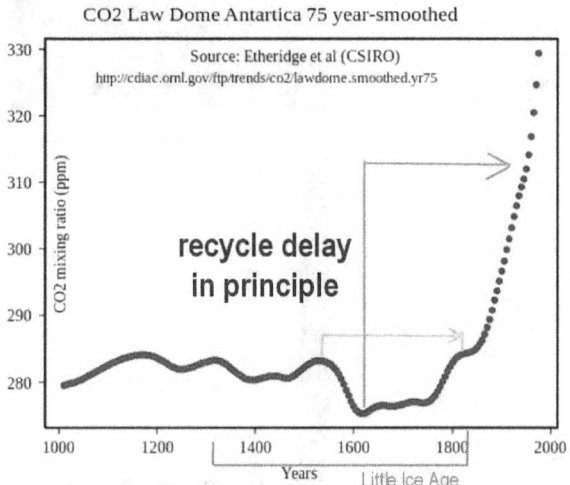

This means that the high rate of CO2 being dissolved into the oceans during the Little Ice Age and earlier cold periods, is coming back to us, into the atmosphere in our time, 350 years later. With this simple scientific recognition in un-constricted science, the noted long recycle delay completely exonerates humanity from the charge of having caused the large increase of atmospheric CO2 that is measured in the modern world.

Here. simple science puts the cause where it belongs, into the court of the natural system, and not into the court of human activity and human energy production. And as I said before, the Primer Fields stand in the background as an element of the electric cosmic process that powers the recycling system.

In the order of a millionth part of it

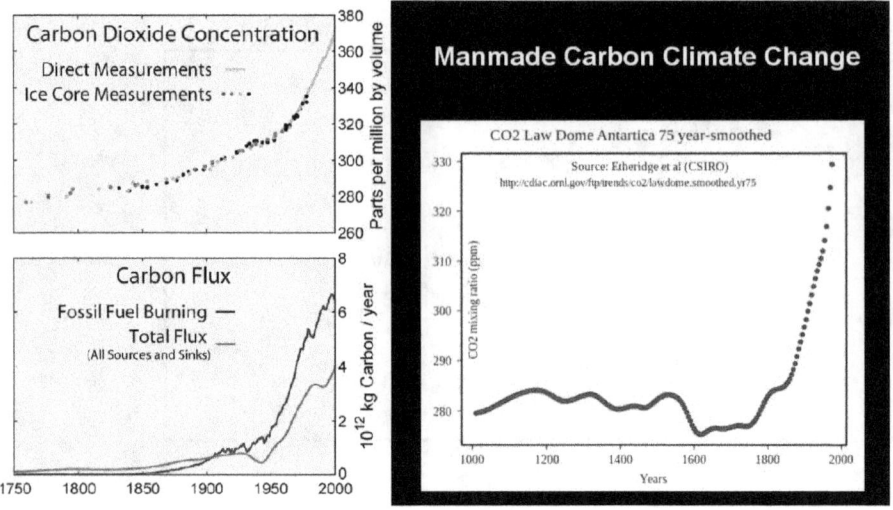

Nor is the CO2 in the air an actual climate factor anyway. It is not physically possible for it to be that. While CO2 is a greenhouse gas that absorbs radiated light energy and reflects it back in a scattered fashion, it is also a scientific fact that its action in the overall context of the greenhouse climate system is extremely minute, in the order of a millionth part of it.

CO2 has no climate-effect

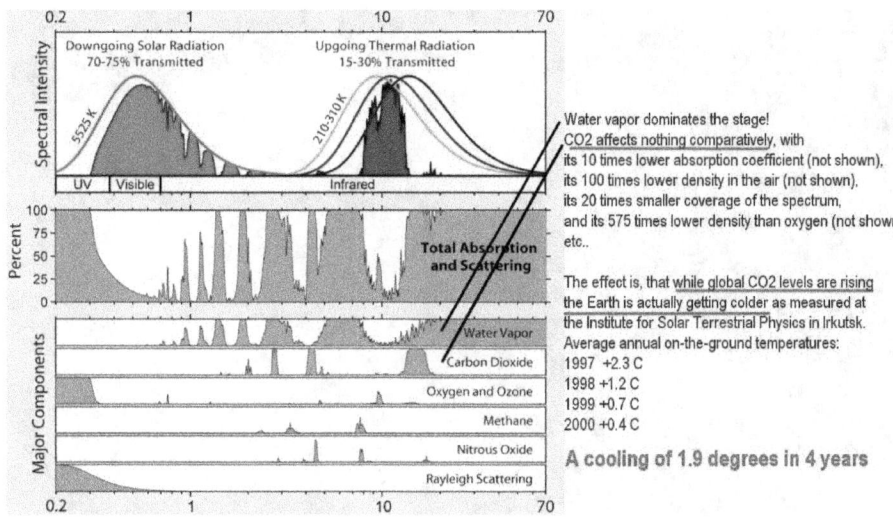

Water vapor dominates the stage!
CO2 affects nothing comparatively, with
its 10 times lower absorption coefficient (not shown),
its 100 times lower density in the air (not shown),
its 20 times smaller coverage of the spectrum,
and its 575 times lower density than oxygen (not shown),
etc..

The effect is, that while global CO2 levels are rising
the Earth is actually getting colder as measured at
the Institute for Solar Terrestrial Physics in Irkutsk.
Average annual on-the-ground temperatures:
1997 +2.3 C
1998 +1.2 C
1999 +0.7 C
2000 +0.4 C

A cooling of 1.9 degrees in 4 years

For all practical purposes CO2 has no climate-effect, and certainly not one that is actually measurable. Its absorption coefficient is 10 times weaker than that of water vapor, which is 100 times more densely present in the atmosphere and is 20 times more widely responsive across the spectrum. And all this is vastly overshadowed by oxygen with an absorption coefficient as high as water vapor, but with a 5-fold greater density in the air. And not less significant is the Raleigh Scattering effects of oxygen and nitrogen in the air that adds to the greenhouse radiation. CO2 adds up to nothing in comparison.

Constricted science sometimes claims that CO2 captures outgoing energy radiated back from the Earth. It is not acknowledged however, that outgoing radiation contributes only about 9% of the atmosphere's total heat budget. But there too, CO2 is vastly overshadowed by water vapor.

Most of the outgoing energy

ISS-34 - Stratocumulus clouds

Most of the outgoing energy has actually nothing to do with greenhouse gases at all, not even with water vapor. The largest climate factor is the effect of cloudiness. The white top of clouds directly reflects a portion of the incoming solar energy back into space, which is thereby lost to us. The amount of cloudiness has a huge effect on the climate.

Latent heat that is released

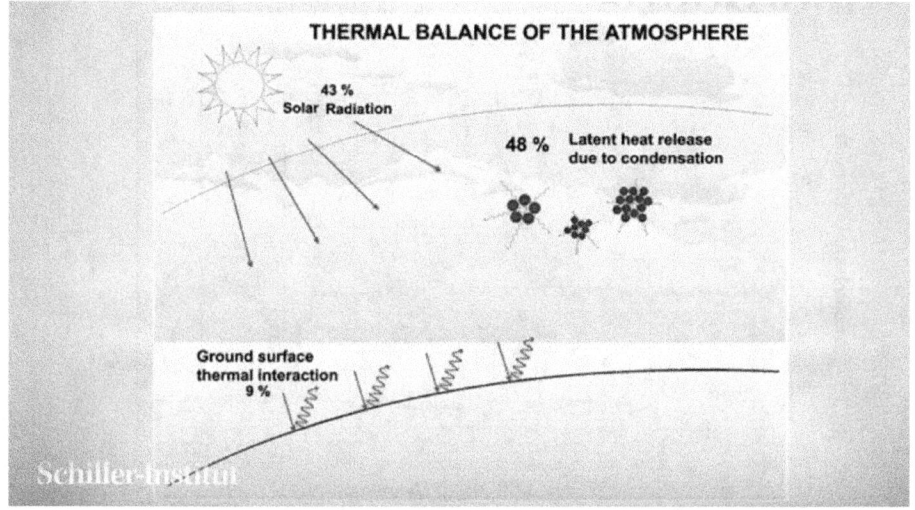

Cloudiness also has a large effect on the atmospheric heat budget. Slightly less than half of the atmospheric heat budget is supplied by latent heat that is released when water vapor condenses back into liquid droplets. This heat is generated in the clouds, high in the atmosphere, and is to a large degree cooled off into the colder region above the clouds.

The Primer Fields stand in the background of the process that largely affects and controls cloudiness on Earth, as the biggest climate factor. CO2, in comparison, adds up to nothing. Its effect is that minuscule. It doesn't even enter the scene where effects are measurable.

Even if the CO2 density in the air was increased 2000%, the increase would have no measurable effect on the climate. A 2000% increase of a millionth part still doesn't add up to anything significant. The increase would still remain below what is measurable in the context of the total climate effects.

That's where CO2 is a factor

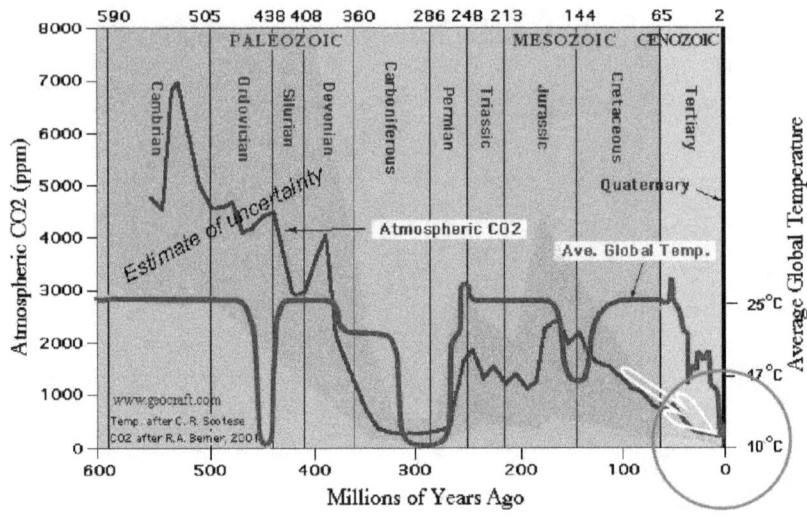

The 2000% increase, which adds up to a 20-fold increase of the global atmospheric CO2, would of course become supremely measurable in the arena where CO2 does have a measurable effect. We presently live in a severely CO2-starved world, with the lowest atmospheric concentration in hundredth of millions of years. The world is presently so severely CO2 starved, that when greenhouse operators increase the CO2 density in their facilities 2-fold, a 50% increase in plant-growth results. Imagine what a 20-fold increase would accomplish, of the type that we had in historic times, or even just 10-fold, as we had it in the Jurassic age, the world would become wonderfully green again, with richer harvests for human living. That's where CO2 is a factor. It's not a climate factor. It simply isn't.

The solar-forced Great Global Cooling

ISS-34 - Stratocumulus clouds

Cloudiness is the big, and hugely variable climate factor.
The solar-forced Great Global Cooling has begun.

Part of the system that affects cloudiness on Earth

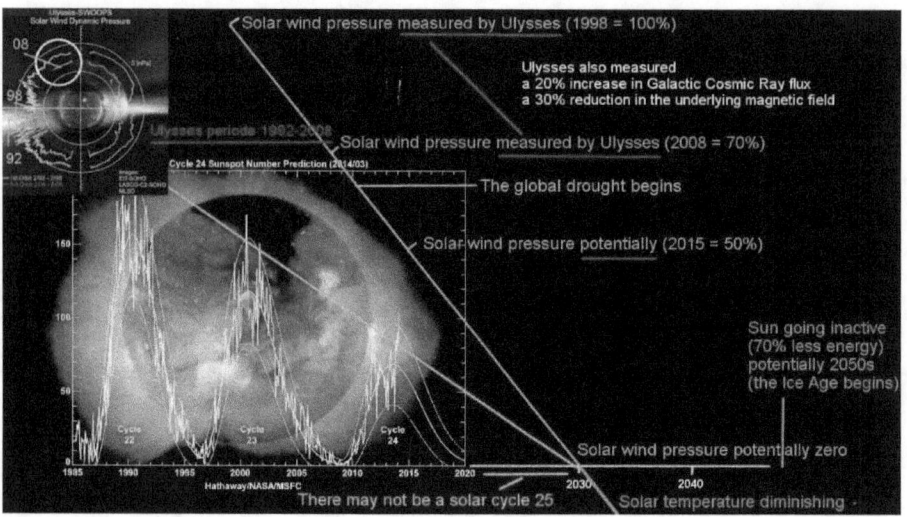

Since the Primer Fields are a part of the system that affects cloudiness on Earth, the subject of the Primer Fields is an important subject in science to be considered, even on the smaller scene that is not directly related to Ice Age dynamics, which, however are ultimately linked to it.

The wide range of considerations that come into focus here, in the context of the Primer Fields, of which the dimming Sun is of course the ultimate concern, evidently determines the wide range of topics that this video series must focus on, and does so.

A scientist is an economist

Topics:

* Introduction to the dimming Sun
* Effects of the Primer Fields on the Sun
* A digital Ice Age
* The Primer Fields dynamics
* Dansgaard Oeschger oscillations
* The UFO phenomenon
* Orbit dynamics of the planets
* The dawn of humanity
* Politics versus science

The video series focuses on the theories and the evidences, and how they will impact us all when the changing conditions unfold as they are scientifically understood to unfold by the nature of the principles involved. In the final video of the series, I will also focus in retrospect on the dimension of our response to the known conditions that affect our future, which are known by their principles. The human response factor to what is fully known, remains presently the biggest open question of them all. Will we become true to ourselves as human beings - true to what we know - and direct our future with what we know? the answer is critical for the very existence of humanity in the near future. It also has the potential to be an open door to the brightest, scientific, technological, economic, and cultural renaissance of all times.
As I said in one of my novels, in essence: a scientist is an economist. We utilize science to protect and enrich our world for the wider benefit of humanity as a whole, which, invariably is to our own benefit.

This means that economists should also be scientists, which may be required in the future, even for politicians as an entrance criterion.

The Effects of the Primer Fields on the Sun

The plasma environment in our solar system

David LaPoint - The Primer Fields

The shape of the galaxy shown here illustrates to some degree, in principle, the shape of the plasma environment in our solar system, though the density in the solar system is presently too low for the organizing plasma streams to be seen.

In the laboratory environment

A plasma sun born in the laboratory

David LaPoint - The Primer Fields

However, by replicating the shaping process in the very small, in the laboratory environment, we can explore the forces that shape these types of phenomena.

David LaPoint discovered in laboratory experiments

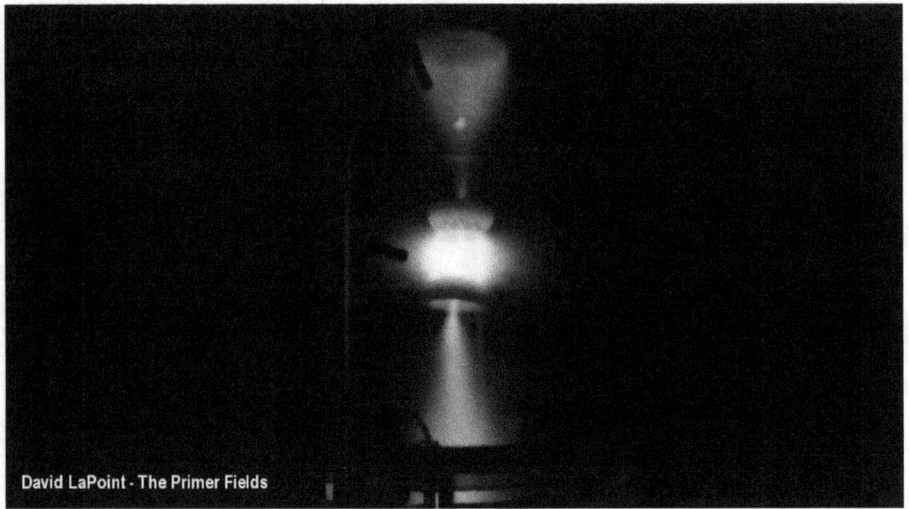

David LaPoint - The Primer Fields

David LaPoint discovered in laboratory experiments that the Primer Fields, when they exist and are functioning, physically prime the environment of the solar system with a densely compressed sphere of plasma centered on our Sun.
The principle is illustrated here.

The effects are amazingly critical

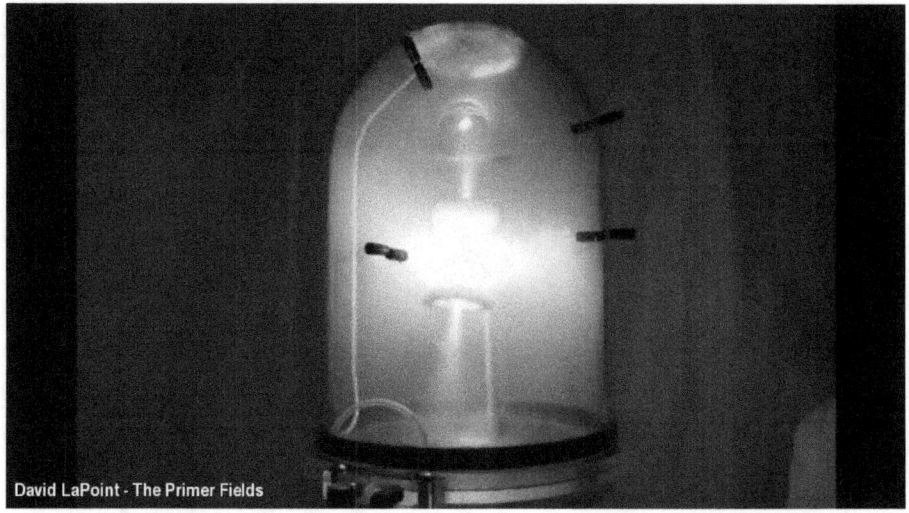

David LaPoint - The Primer Fields

The effects, however, that David LaPoint discovered, are so amazingly critical that great problems arise for humanity when their functioning is impeded or is collapsing.

Condensed plasma interacts with the Sun

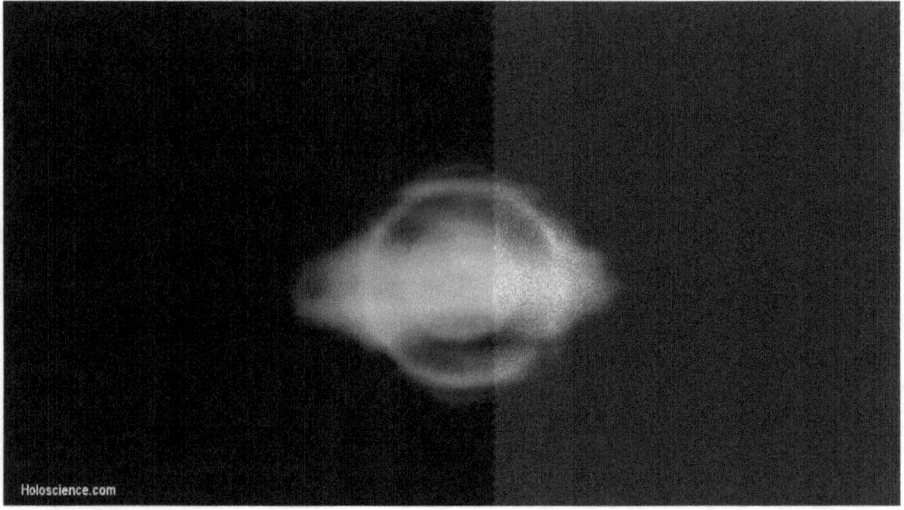

The evidence tells us that the dense plasma environment that is focused on the center of the solar system enables our Sun to be electrically powered from the outside by electric arc reactions. In these reactions the condensed plasma interacts with the Sun's photosphere.

Illustrated in the Red Square nebula

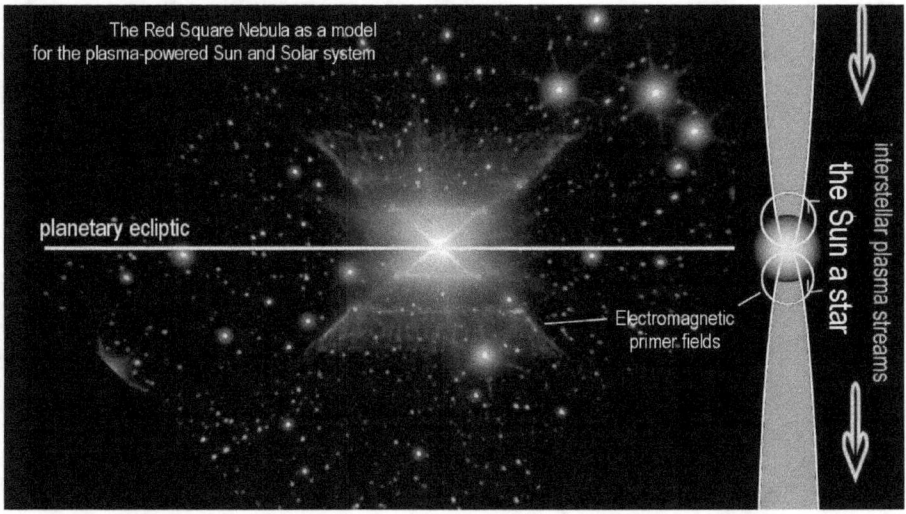

The Red Square Nebula as a model for the plasma-powered Sun and Solar system

planetary ecliptic

Electromagnetic primer fields

the Sun a star

interstellar plasma streams

The condensing process is a multi-stage process that is illustrated to some degree in the Red Square nebula. The nebula is not the remnant of an exploding star as nebulas are often regarded. Instead it is an example of the typical features of the Primer Fields in operation. The fields form by the principles inherent in the flow of electric plasma in space. The details will be discussed later, though the overall effect is noteworthy here. The effect of the fields is that plasma streams that exist in galactic space in wide channels, becomes drawn together, like by a wide funnel, and becomes magnetically focused from the funnel onto a central sun, or a system of multiples suns that become intensely powered by this process. The planets exist outside of this densely powered sphere, on an ecliptic centred on the Sun, between the two Primer Field structures. The plasma concentration process that we see in operation here, which is basically the same for every sun, renders our Sun as an intensely powered catalytic energy converter, that is powered not from within, but is powered externally at its surface.

71

The Sun's great brilliance generated at its very surface

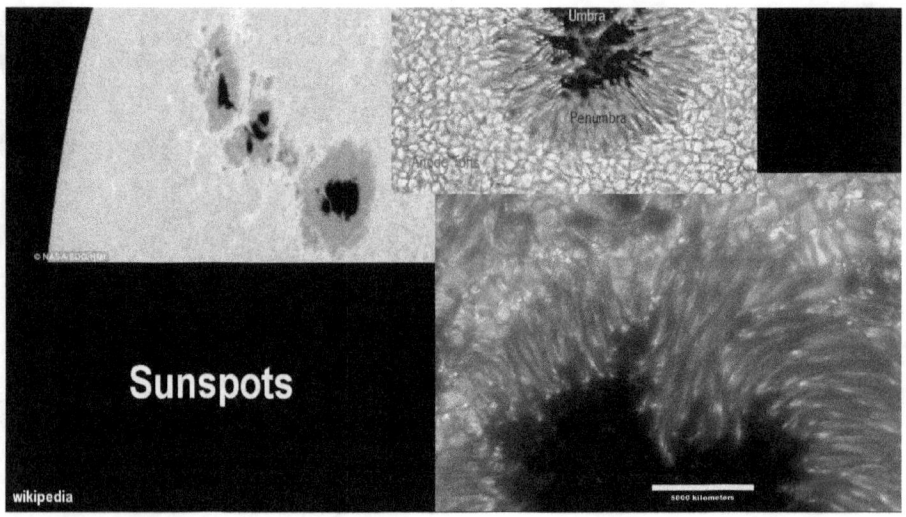

The evidence that the Sun is externally powered is fairly obvious. When we look beneath the high-powered photosphere, through the open space at the umbra of the sunspots, the Sun reveals itself as being much darker, and therefore cooler inside. The photosphere has been measured at a whopping 5,870 degrees Kelvin, and the surface below at a mere 3,000 degrees. This means that the Sun's great brilliance is electrically generated at its very surface and does not emanate from within from nuclear fusion reactions as it is generally believed.

The plasma concentration process

David LaPoint - The Primer Fields

The plasma concentration process that enables a sun to be electrically powered, has been replicated in lab experiments.

A highly compressed plasma sphere was formed

David LaPoint - The Primer Fields

There a highly compressed plasma sphere was formed in a thinly filled chamber of gas between two bowl-type permanent magnets that replicate the functional elements of the Primer Fields.

Magnetic fields operating environment

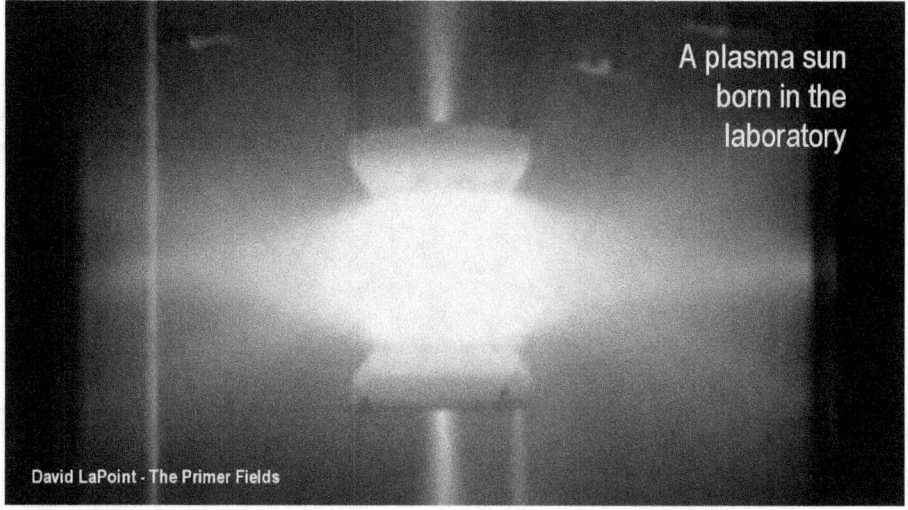

A plasma sun
born in the
laboratory

David LaPoint - The Primer Fields

The result, which is shown here, was develop by the functions of
the magnetic fields acting on the plasma flow to concentrate it.
It is interesting to note that in the lab experiment the plasma
sphere at the center did not form instantly. The operating
environment needed to be established first.

In the real world

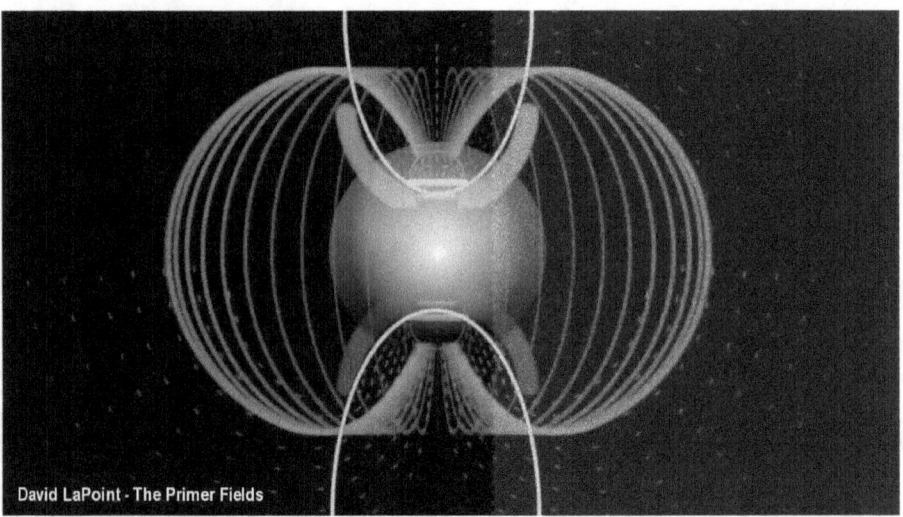

David LaPoint - The Primer Fields

However, in the real world this establishing-function also includes the forming of the bowl-shaped magnetic fields themselves by which the plasma becomes concentrated. In the lab, the magnetic fields were provided with manufactured magnets. In the real world the bowl-shaped magnetic fields are electrically created structures that are formed by moving electricity in plasma streams.

The plasma sphere in which our sun is located

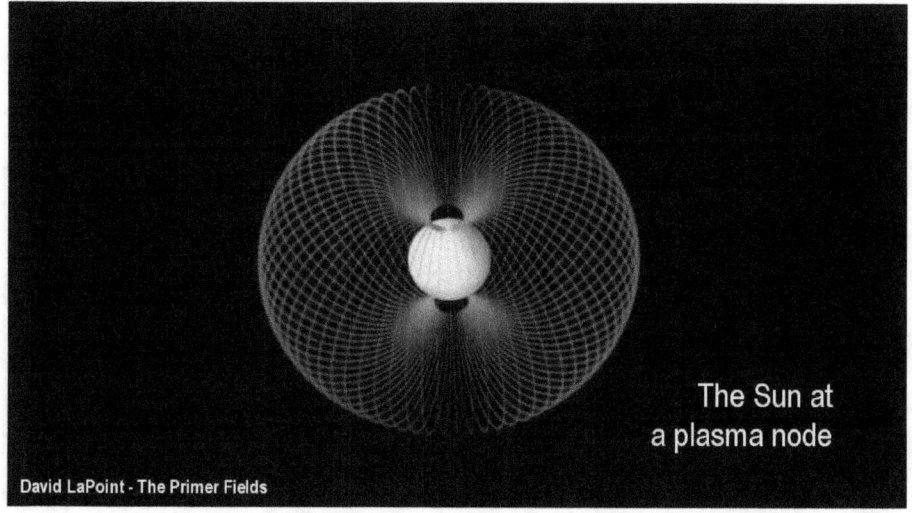

The Sun at
a plasma node

David LaPoint - The Primer Fields

This means that the plasma sphere in which our sun is located, is the composite result of complex processes working together, for which a certain level of plasma input density is required for the forming of the magnetic bowl structures to happen.

I will illustrate later how the bowl-shaped magnetic structures are formed, that furnish the Primer Fields, which in turn create the conditions that focus plasma onto our Sun that lights it up to great brilliance.

Different types of atoms emit light in different bands

The spectrum of sunlight

While different types of atoms emit light in different bands, the numerous types of atoms that are created in the photosphere, in combination, create photons of all possible shapes and sizes that together cover the entire light spectrum and beyond, in a seamless field of colors.

The color-rich world that we cherish

The white sunlight that we see reflected in the color-rich world that we cherish, is the direct result of the plasma-fusion process in the photosphere of the Sun that is facilitated by the actions of the Primer Fields that focus interstellar plasma unto the Sun. The white sunlight with its colorful spectrum is not possible on any other basis.

The Sun as a sphere of hydrogen gas

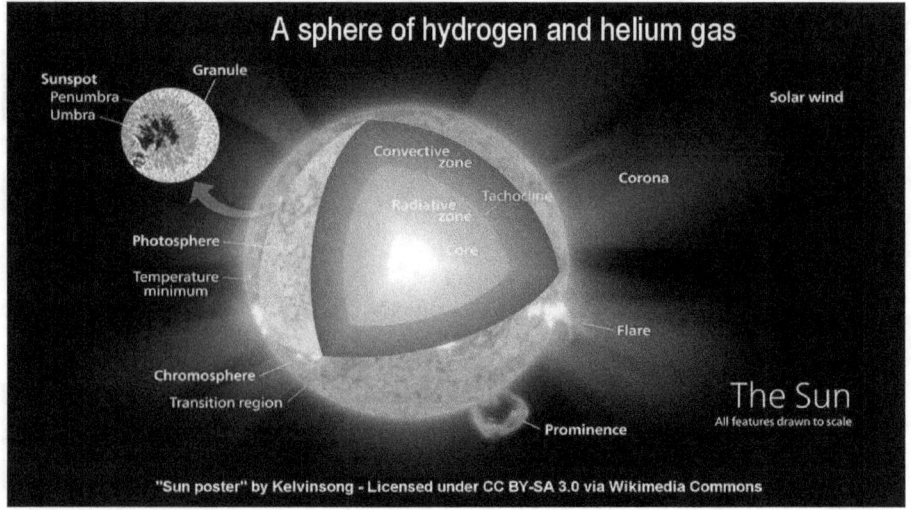

The historically theorized concept of the Sun as a sphere of hydrogen gas with a hydrogen-fusion process at its core, is obviously false, because it cannot produce the white sunlight spectrum that we see.

Hydrogen atoms emit light in only a few narrow bands

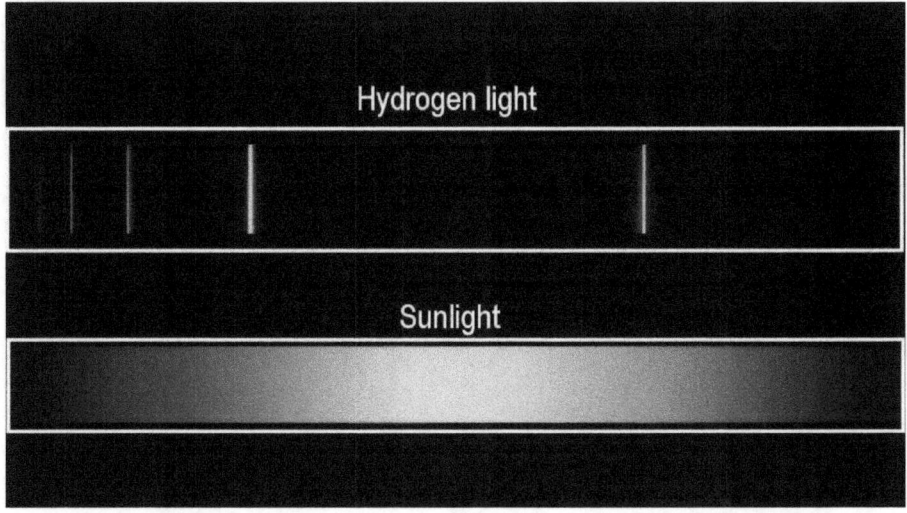

Hydrogen atoms emit light in only a few narrow bands.
The white sunlight spectrum is only possible by solar surface plasma
fusion where all known types of atomic elements are created.

The color-rich white sunlight is clear tangible evidence

The color-rich white sunlight is clear tangible evidence that we live in a solar system powered by plasma streams that the Primer Fields are an active element of, centered on a plasma Sun.

The surface plasma fusion also emits highly energetic solar cosmic-ray flux

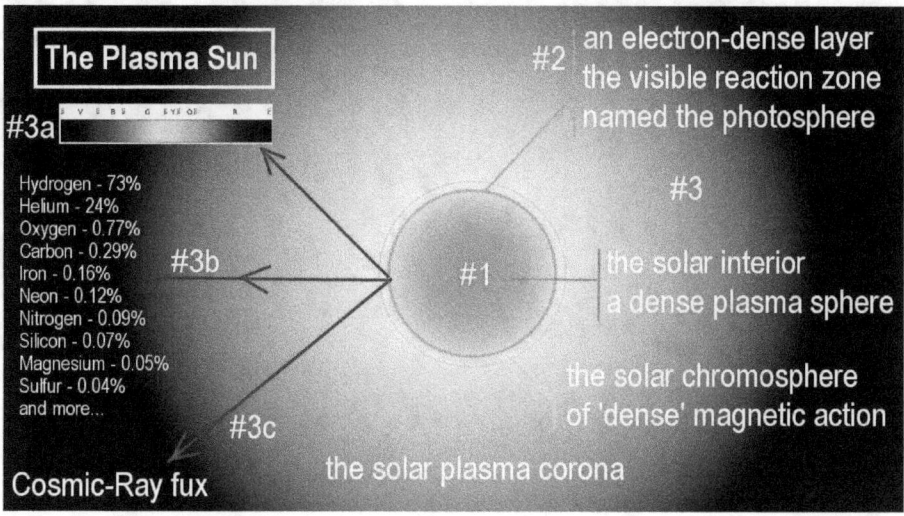

The Plasma Sun

#3a

Hydrogen - 73%
Helium - 24%
Oxygen - 0.77%
Carbon - 0.29%
Iron - 0.16%
Neon - 0.12%
Nitrogen - 0.09%
Silicon - 0.07%
Magnesium - 0.05%
Sulfur - 0.04%
and more...

#3b

Cosmic-Ray fux

#3c

the solar plasma corona

#2 an electron-dense layer
the visible reaction zone
named the photosphere

#3

#1

the solar interior
a dense plasma sphere

the solar chromosphere
of 'dense' magnetic action

However, the surface plasma fusion does not only create atomic elements and streams of light, but also emits highly energetic solar cosmic-ray flux, marked 3C. Cosmic rays are not streams of light, but are events of highly energized individual electrons and protons escaping from the magnetic confinement of the primer fields of the plasma fusion cells in the photosphere. Most of the solar cosmic rays are trapped in the corona, marked #3, but when the corona weakens, more of them penetrate the barrier and reach the Earth.

*Solar cosmic rays have an ionizing effect

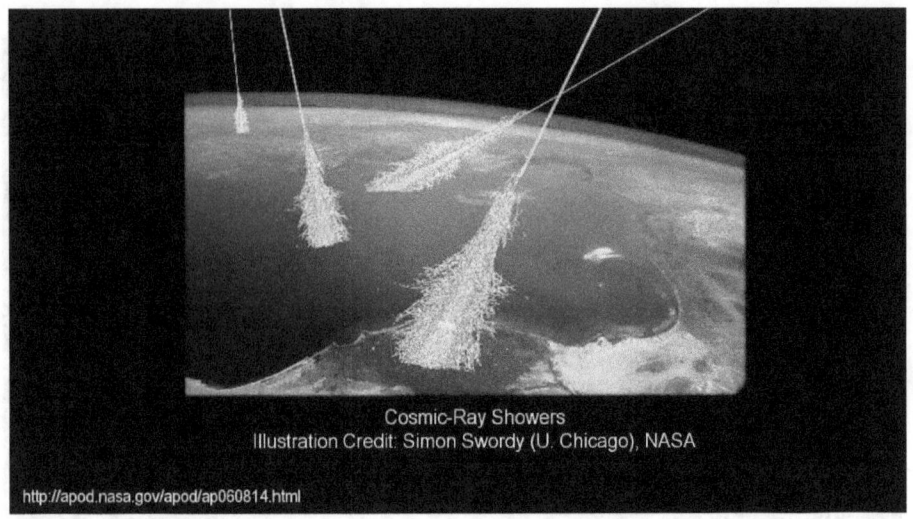

Cosmic-Ray Showers
Illustration Credit: Simon Swordy (U. Chicago), NASA

http://apod.nasa.gov/apod/ap060814.html

Solar cosmic rays have an ionizing effect in the atmosphere that enhances the cloud forming process, which affects the climate on Earth.

Increased cloudiness results in colder climates

Increased cloudiness results in colder climates. The white top of the clouds reflects a portion of the incoming solar energy back into space, which thereby becomes lost to us. The top of clouds also radiates latent energy from the cloud-forming process into space. Latent energy is released when water vapor is condensed into liquid droplets.

When a pot of water is boiled into steam

Maximum temperature of liquid water at ambient pressure is 100 degrees Celsius: The Boiling Point

When a pot of water is put on a stove and is boiled into steam, the energy that is invested in the process is released as latent energy when the steam condenses back into water.

Clouds cooling latent energy into space

In the atmosphere, 45% of the thermal budget is derived from latent energy. Much of this energy is released in the clouds, at the edge of the atmosphere, where much of it is cooled into space. By the combined effect of clouds reflecting solar energy directly back into space, and the clouds cooling latent energy into space, the rate of cloud forming is the major climate-determining factor on Earth, and the rate of this process is absolutely determined by the Sun, by the volume of cosmic-ray flux that escapes from it.

When the solar corona is weak

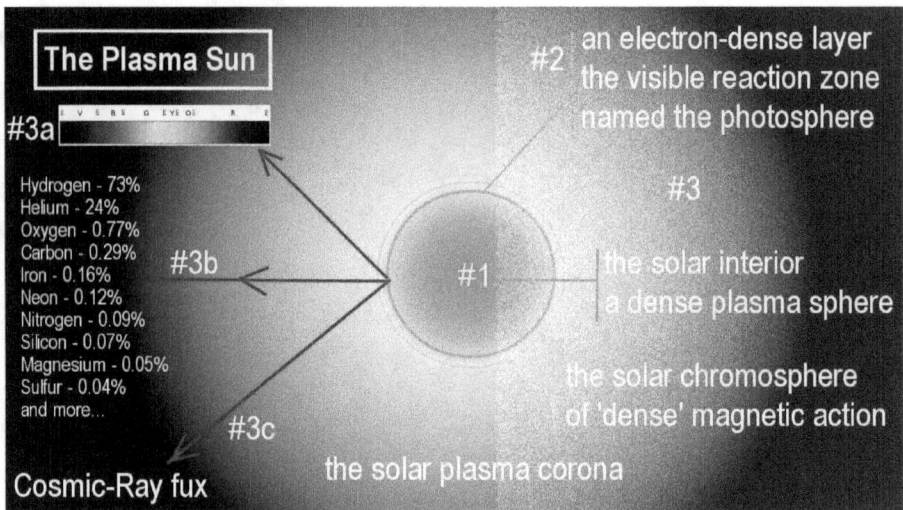

When the solar corona is weak, which also results in weak solar activity, the solar cosmic-ray flux is increased, cloudiness is thereby increased, and the Earth gets colder. When this happens in a big way, a Little Ice Age results.

That's what we saw in the 1600s

That's what we saw in the 1600s, at the time of the Maunder Minimum of the solar activity. The global cooling was so massive that rivers became skating rinks in the winter.

Indicated in Carbon-14 measurements

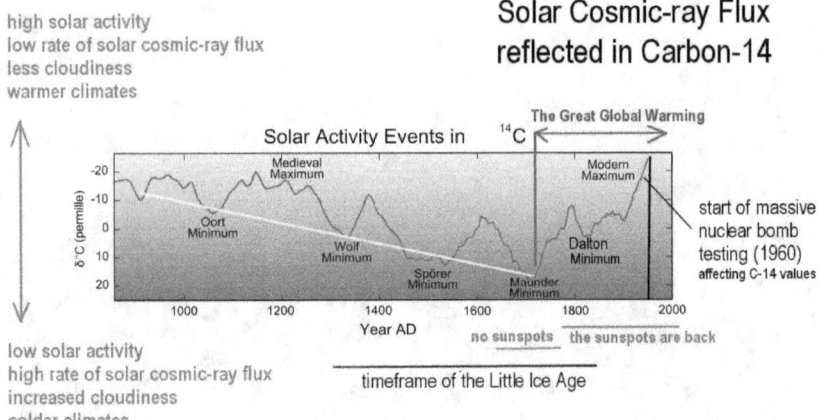

That this time of the Little Ice Age was a time of large volumes of solar cosmic-ray flux, is indicated in Carbon-14 measurements. Carbon-14 results from Solar cosmic rays affecting the atmosphere. All the cold periods that we have measurements for were periods of weak solar conditions that result in extremely high rates of solar cosmic-ray flux, and by implication also high rates of cloudiness.

The ever-changing climate on Earth

This means that the ever-changing climate on Earth, is directly caused by changing density in the interstellar plasma stream, which, through the Primer Fields system, is focused onto our Sun. It also proves that the Sun is surface-powered by the Primer Fields, because no other platform than surface plasma fusion is able to generate solar cosmic-ray flux, and this so massively that it affects the climate on Earth.

Global warming after the Little Ice Age, was not manmade

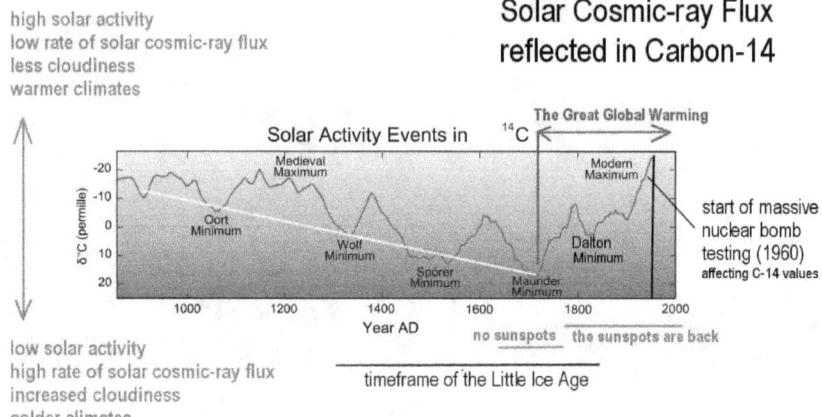

It also means that the great global warming after the Little Ice Age, was not manmade by industrial activity and fuel burning, but was the direct result in changing cosmic conditions that the Primer Fields focused onto the Sun where it affected the Sun's delicate operating dynamics.

The recognition of these changing cosmic conditions is critical for our time, because we see in them a long-term down-ramping in progress towards another Little Ice Age coming up fast, from which the solar system may not recover.

The solar system barely recovered from the Little Ice Age

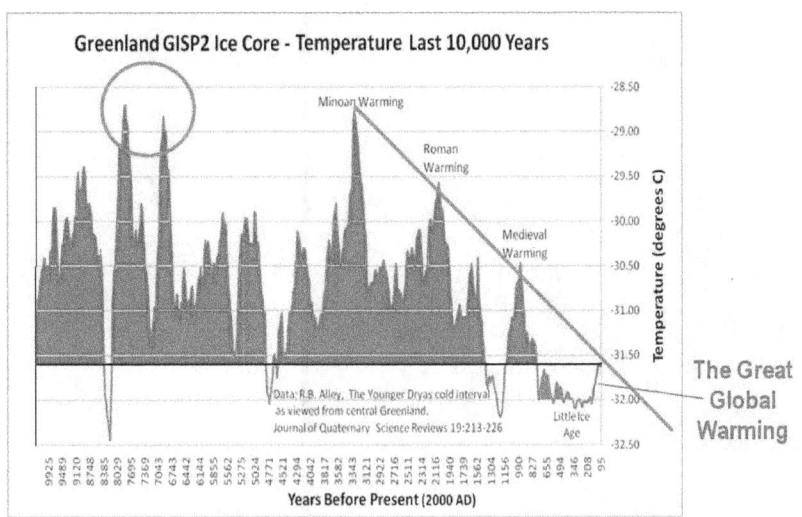

The solar system barely recovered from the Little Ice Age. The down-ramping that we see in suggests that the next minimum in the solar system may take us below the minimal density that is needed for the Primer Fields to be maintained.

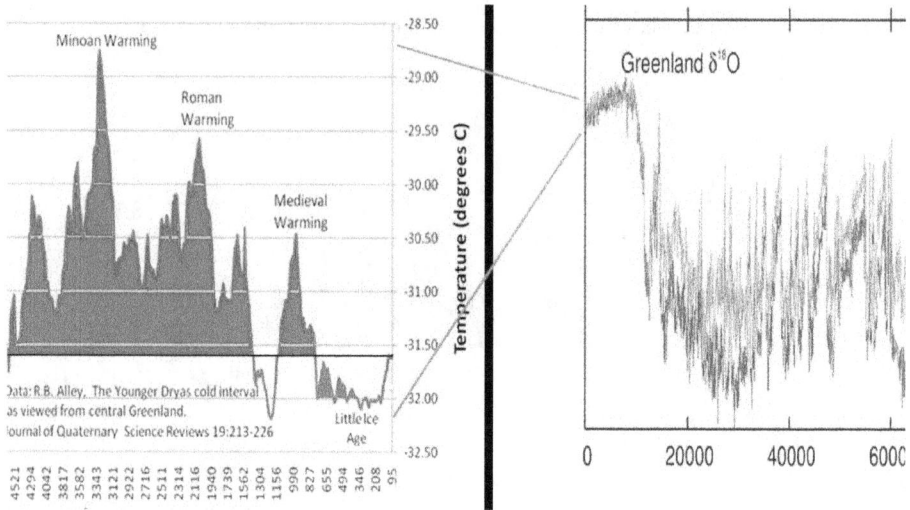

When the Primer Fields collapse, a phase shift occurs. Then all the climate changes that have occurred in the last 10,000 years will appear as nothing in comparison. With the Sun going largely inactive, the glacial climate begins that ice core measurements tell us, will be 40 times colder than the worst of the Little Ice Age had been.

It has become evident in lab experiments

It has become evident in lab experiments and cosmic observations that when the input density of the plasma streams that feed into the Primer Fields system drops below a certain minimum threshold, the critical fields don't form, whereby the entire process stops.

The solar system is not as robust as is generally believed

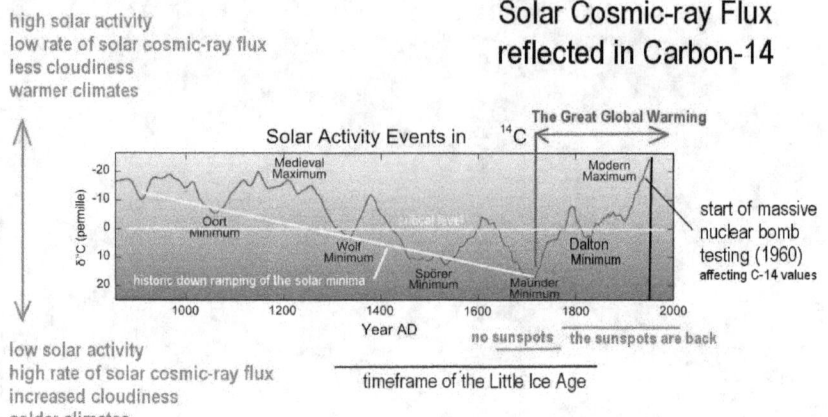

"Carbon14 with activity labels" by Leland McInnes at the English language Wikipedia. Licensed under CC BY-SA 3.0 via Commons

The climate fluctuations that have been experienced, that we have records of, tell us that the solar system is not as robust as is generally believed, and is down-ramping.

A colder, darker, yellow sun

When the phase-shift happens in the real world, the brilliant photosphere of our sun is no longer being powered. It becomes in effect, turned off. It fades into a thin haze, leaving in the wake a colder, darker, yellow sun that glows dimly by its internally stored-up energy like ambers of a fire gone out, which too, then slowly diminish.
This means that the transition to the next ice age will not be a slow process that is drawn out over thousands of years, but will be a radical turn-off transition in which the Sun goes dim in possibly a single day.

A deactivated Sun

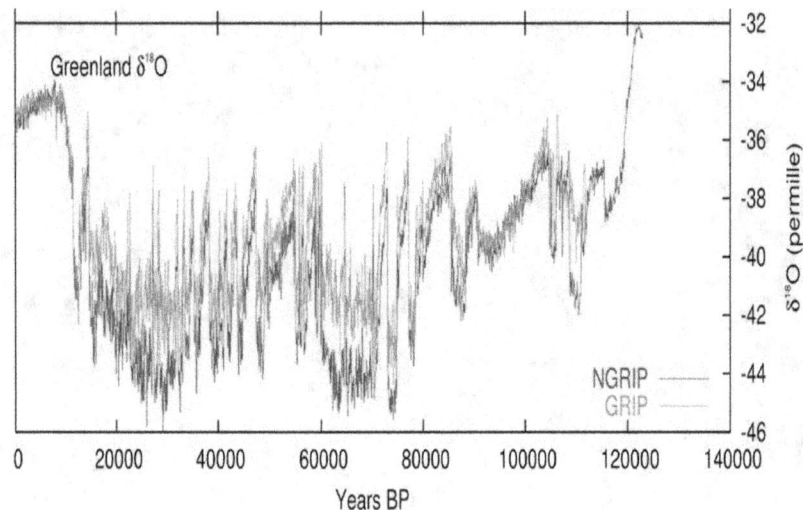

There exists plenty of evidence in ice core samples that the kind of rapid transformation of the climate on earth has happened that reflects an on-off transition, from an active Sun, to a deactivated Sun.

Nothing short of such a radical transition can explain the massive, rapid cooling that occurred when the last Ice Age began 120,000 years ago.

A massive reduction in solar energy

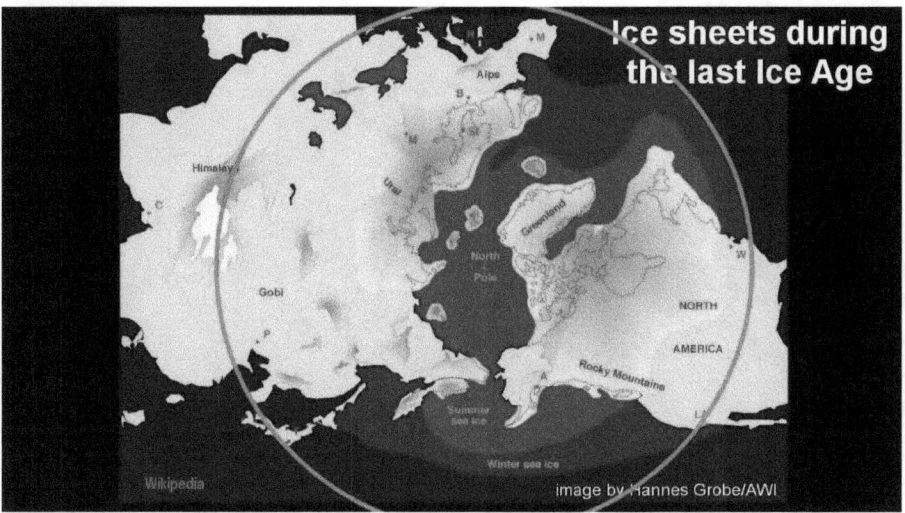

It takes a massive reduction in solar energy input into the Earth's climate, to cause the formation of the enormous ice sheets that become spread across much of the northern hemisphere as it is shown here, which piled up 10,000 feet thick, or more in some places.

A radically different world unfolds

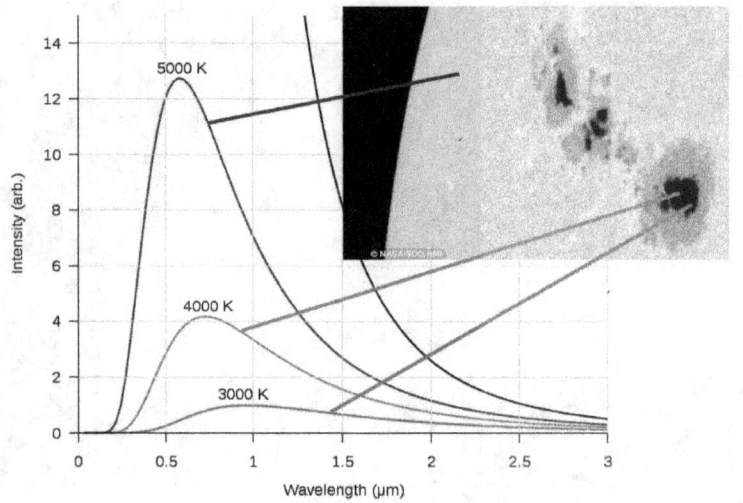

When the phase-shift happens a radically different world unfolds. The actual timeframe in which the phase-shift unfolds may be extremely short, probably spanning less than a single year for all practical purposes.

We are presently near the phase-shift point

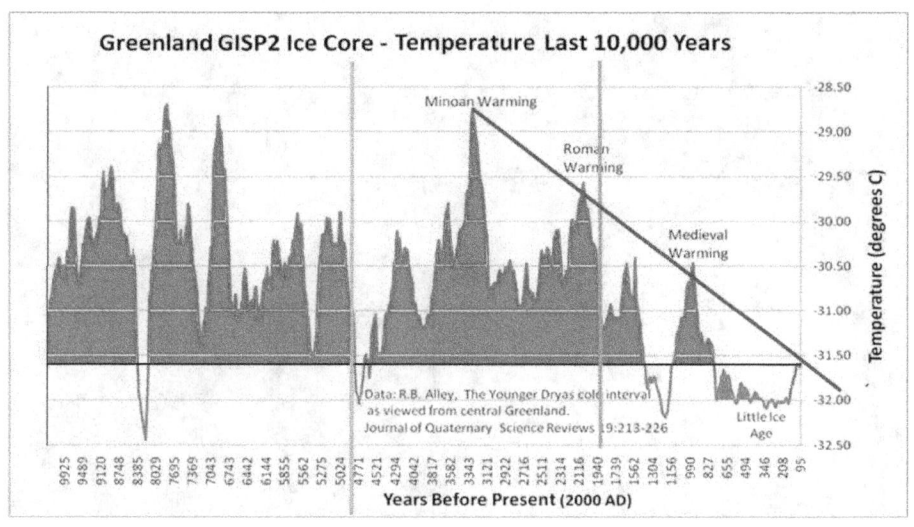

We are presently near the phase-shift point. We have come to the end phase of the actively powered period of the sun, and the beginning of the next 90,000-year glaciation period.

Of course we don't know the precise day or year at which the Primer Fields will collapse and the plasma sphere around the Sun will vanish, but do we really need to know this?

We know already enough to be inspired by it to take the appropriate actions to build new infrastructures for our food supply and for our living. We know with measured evidence on hand that the Earth has been in a diminishing trend for 3,000 years already, that appears to be now accelerating towards the critical cut-off point.

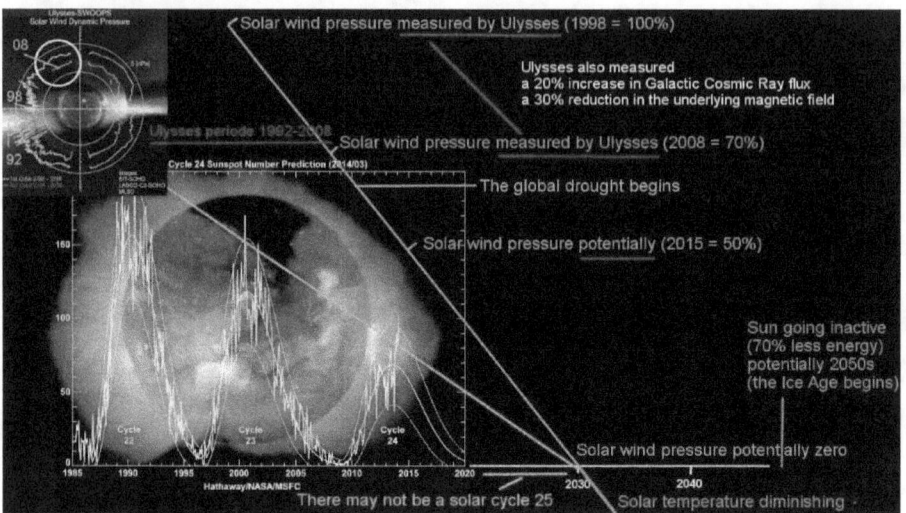

We see the solar system getting weaker. The tell-tale sunspot cycles are diminishing. Also, over the timeframe of a single decade, NASA's Ulysses spacecraft has measured an amazing weakening of the solar wind pressure by a whopping 30%. This is an enormous drop-off for such a short period.

When the Sun's powered state ends

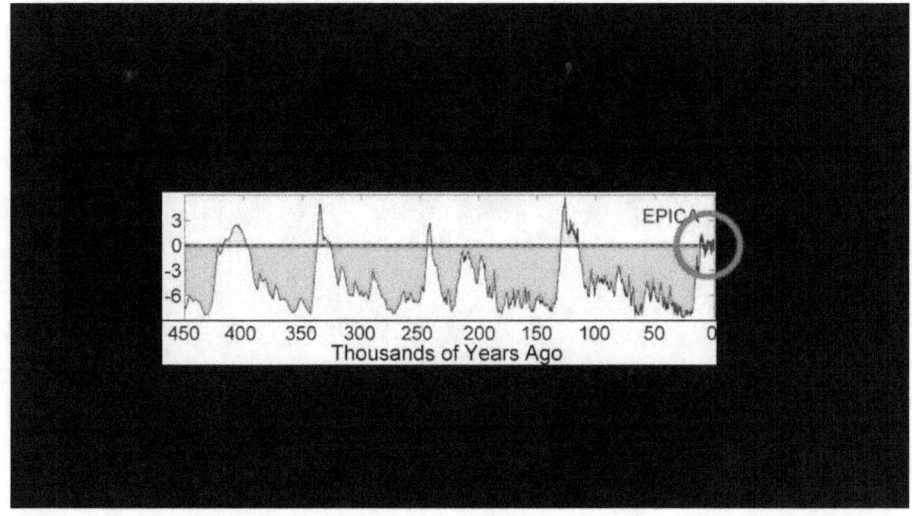

We also know that when the Sun's powered state ends, that is when it gets turned off to a lower intensity state, the big ice sheets will form. It takes a big change in the incoming energy that warms our planet, for big effects to happen.

No one is prepared for the consequences

photo by Andrew Mandemaker

Let's hope that the cut-off point that we are moving towards is still more distant in time than it appears to be, because at the present time no one of humanity is prepared for the consequences.

The Ice Age consequences promise to be far bigger

Not the least preparations are even considered to be made, much less are made, even while the Ice Age consequences, when they begin, promise to be far bigger than most people dare to imagine.